Disaster Research and Management Series on the Global South

Series Editor
Amita Singh
Centre for the Study of Law and Governance & Special Centre for Disaster Research
Jawaharlal Nehru University
New Delhi, India

Disaster Research and Management Series on the Global South is a series coming out of Special Centre for Disaster Research (SCDR) at Jawaharlal Nehru University (JNU), Delhi, India. SCDR is the first in Asia Pacific to start a course on disaster research within a social science perspective. The series follows and publishes pedagogical and methodological change within the subject. The new direction of teaching, research and training turns from 'hazard based' to 'resilience building'. The series taps such research for the benefit of institutes and higher education bodies of the global south. It also suggests that much of the western literature based upon rescue, relief and rehabilitation which is also being taught in the Asian institutes is not directly relevant to managing disasters in the region. It provides reading and study material for the developing field of disaster research and management.

1. Generates a non-west transdisciplinary literature on disaster research and studies
2. Strengthens disaster governance and improves its legal framework
3. Sensitizes disaster management authorities towards key priorities and attention areas
4. Focus on preparedness is strongly proposed and revisited
5. Highlights changes in pedagogy and methodology of disaster research and teaching
6. Mainstream vulnerable communities of differently abled, elderly, women, children
7. Indicate strategies needed to protect city animals, birds and wildlife during disasters

More information about this series at
http://www.palgrave.com/gp/series/16402

T. V. Vijay Kumar • Keshav Sud
Editors

AI and Robotics in Disaster Studies

palgrave
macmillan

Editors
T. V. Vijay Kumar
School of Computer and Systems Sciences
Jawaharlal Nehru University
New Delhi, India

Keshav Sud
Data Science
Keurig Dr Pepper
Burlington, MA, USA

ISSN 2662-4176 ISSN 2662-4184 (electronic)
Disaster Research and Management Series on the Global South
ISBN 978-981-15-4290-9 ISBN 978-981-15-4291-6 (eBook)
https://doi.org/10.1007/978-981-15-4291-6

This Palgrave Macmillan imprint is published by the registered company Springer Nature Singapore Pte Ltd.
The registered company address is: 152 Beach Road, #21-01/04 Gateway East, Singapore 189721, Singapore

Acknowledgements

Our deepest gratitude to none other than our former Minister of State for Disaster Management Hon'ble Shri Kiren Rijiju who inspired, encouraged and supported research into new areas of disaster management. His enthusiasm, office room discussions, his encouragement to pursue unattended areas in disaster research and his unending ambition to make India the best in the world in implementing the Sendai Framework pushed us to deeply focused research. He is now adorning the sports ministry but his contribution to the setting up of India's first completely research-based disaster research centre at Jawaharlal Nehru University (JNU) has been the most illuminating feather in his cap. We hope that his contribution reaches its desired destination from a sapling to a full grown tree.

Some officials from administrative services have joined hands with us in our endeavour like the former Executive Director of the National Institute of Disaster Management (NIDM) Mr. Anil Kumar and the current Executive Director Major Gen. Manoj Kumar Bindal for enabling a wholesome participation, providing an atmosphere of academic freedom and enthusing empirical research in the NIDM faculty. The faculty of the Special Centre for Disaster Research (SCDR) at JNU and the NIDM could collectively encounter challenges at the field and contribute to a variety of curriculum-building events at JNU due to the valuable support coming from the former Minister of State for Home Affairs Shri Kiren Rijiju. Much research for the book has actually taken place during his tenure as the Minister of State for Home Affairs. It is also worth mentioning here that on 8 August 2015 Shri Rijiju Ji was instrumental in getting

NIDM and JNU sign an MoU for collaboration on disaster research and three months later on his birthday, 19 November, he was designated by the UN Office for Disaster Risk Reduction (UNISDR) as "Disaster Risk Reduction Champion for the Asia Region". The JNU Research Team on Disaster Research recollects the attention, focus and clarity with which he corrected research teams of seven professors with respect in his office, sometimes a virtual classroom.

SCDR is equally appreciative of Lt. Gen. N. C. Marwah, member, National Disaster Management Authority (NDMA) who consistently provided the required backup support during field surveys and data collection in disaster-affected areas. Helping the victims of disasters became his mission and shed light on the studies on administrative leadership in managing disasters.

The work wouldn't have moved further without the constant support of former Registrar Prof. Bupinder Zutshi and the present Registrar Dr. Pramod Kumar in clearing an unending trail of obstructions which came from peer group scholars who had problems understanding the more refreshing transdisciplinary imperatives of social science approach to disaster research displacing the mainstream science- and technology-based approach. The support of Prof. Mondira Dutta (Former Chairperson CIAS, JNU), Prof. Milap Punia (Chairperson, CSRD), Shri Sanjeev Kumar Sharma (Director CIS, JNU) and Dr. Robert Thongkholal Haokip (Faculty CSLG) is unforgettable and is cast in JNU's academic history.

The JNU Central Library deserves special mention. The readiness to help in exploring and providing rare literature and data on disasters came so handy to many authors. Former librarian Dr. Ramesh C. Gaur and after him Dr. Manorama Tripathy understood and appreciated our search for literature on transdisciplinary disaster studies in law, governance, public policy and sciences. One would never forget the vibrancy and intellectual energy of a younger librarian Azmi Khan of a smaller library at the Centre for the Study of Law and Governance at JNU. She almost waited to hear from the research team about their next search and join them in loud youthful discourses.

Our office staff of the Special Centre for Disaster Research, Deepak Kumar, Hemchand Pandey, Darakshan, Jhuman and Narinder, brightened the work with their secretarial assistance and coordination support which helped us immensely in meeting the deadline.

Editors of the book also acknowledge an unforgettable and versatile contribution of the research team led by Gaurika Chugh, Vualzhong Mung, Chetana Attri, and Natasha Goyal who turned this challenge into a delightful exercise of unresolvable debates and never-ending cups of tea and coffee. The long-distance support of Dr. Manika Kamthan and Dr. Binod Kumar was always available like disciplined soldiers of the team.

The most enlightened support arrived from the Springer-Palgrave publishing team guided by the visionary Editorial Director Springer Singapore William Achauer, Associate Editor Palgrave Macmillan Sandeep Kaur and Editor Springer-Nature Nupoor Singh who managed discussions and coordination in an unfailing mode. Their collaboration goes beyond mere publication as they absorbed themselves in the research teams including many workshops, symposiums and debates.

Publishing these volumes of non-west literature has led the SCDR research team to look for many new authors from local administration, affected communities and implementers. The editors acknowledge the shared contribution of many who despite the motivation could not write due to their intensive work responsibilities in the Chennai floods, Cyclone Gaja and Kolkata Bridge collapse. Indian Council of Social Science Research (ICSSR) has empowered many of these implementers who remain knowledge repositories for original literature in disaster studies. The editors appreciate the supportive role of ICSSR in bringing about this volume.

Last but not the least NAPSIPAG (Network of Asia Pacific Institutions of Public Administration and Governance) stands with this initiative strong and determined as ever before. This is one big Asia-Pacific family of policy experts which is always passionate to celebrate collaborations in generating knowledge from their homelands.

Contents

About the Series Editor

Amita Singh Professor at the Special Centre for Law and Governance and Founding Chairperson, Special Centre for Disaster Research, Jawaharlal Nehru University (JNU), India. Professor Singh teaches Law and Governance at the Centre for the Study of Law and Governance at JNU. She has been the longest serving Secretary General of NAPSIPAG (Network of Asia Pacific Schools and Institutes of Public Administration and Governance) initiated by ADB 2004 at INTAN Malaysia. She is Member Secretary of the Institutional Ethics Review Board and Member of the Indian Council of Social Sciences Research, Delhi. She has a wide research experience of evaluating best governance practices and working with the government (DARPG, India) and the Global Innovators Network, at the John F. Kennedy School of Government, Harvard University. She was awarded the Australia-India Council Fellowship (2006–2007) for academic research in nine Australian Universities and was again awarded the Australian Monash Fellowship in 2017. She has been closely associated with the International Women's Association at Hunter College SUNY USA in 1990 to prepare for the Beijing Declaration in 1995. She has received the Bangladesh National Award of 'Nawab Bhadur Syed Nawab Ali Chowdhury National Award' 2014 for women empowerment. Professor Singh is an ardent activist of the 'Rights of Nonhuman Species'.

Notes on Editors

T. V. Vijay Kumar is Professor and Dean at the School of Computer and Systems Sciences, Jawaharlal Nehru University (JNU), New Delhi, India. He is a concurrent faculty at the School of Engineering, JNU and at the Special Centre for Disaster Research, JNU. He is also an associated faculty at the Atal Bihari Vajpayee School of Management and Entrepreneurship, JNU. His publications and research interests revolve around areas such as databases, artificial intelligence, machine learning, nature inspired algorithms, disaster management and big data analytics.

Keshav Sud obtained his PhD from the University of Illinois at Chicago. He has worked for Fortune 500 companies like Amazon, Caterpillar, Volvo, and currently holds the role of Senior Manager of Data Science at Keurig Dr Pepper Inc. He is leading the development of recommendation engines. He has authored many publications in internationally recognized professional journals. He serves on the editorial board of many American and Asian journals. He recently published the book *Introduction to Data Science and Machine Learning* (co-eds. Pakize Erdogmus and Seifedine Kadry)

List of Figures

List of Tables

CHAPTER 1

Introduction: Enhancing Capacity to Manage Disasters

Amita Singh

Introduction

The impact of ICT on the functioning of governance institutions has come to a stage where some immediate and comprehensive steps should be taken. More than two and a half quintillion of data is produced every day in the world and 90 percent of all data today has been produced in the last two years. This indicates that governance is likely to get buried or become irrelevant under the load of data. This directs attention towards the problem of organizing data. Big Data (BD) suggests that even digitization of information has reached its saturation point and is now to be stored through higher analytical skills in governance. These special skills are required for use in identifying content as well as their analytical relevance which could be used later or whenever required. Google's Eric

A section of this paper was published in the Sri Lankan Journal of Business Economics (Vol. 4, 2013) as 'Enhancing Capacity to Govern Through Big Data' by the author. The current paper is an updated version.

A. Singh (✉)
Centre for the Study of Law, Governance & Disaster Research, Jawaharlal Nehru University, New Delhi, India

T. V. V. Kumar, K. Sud (eds.), *AI and Robotics in Disaster Studies*, Disaster Research and Management Series on the Global South, https://doi.org/10.1007/978-981-15-4291-6_1

Schmidt writes, "Our propensity for selective memory allows us to adopt new habits quickly and forget the ways we did things before." (p. 8). BD enables us to keep track and simplify the crowding of scattered data which is creating a 'data tsunami' with the communication companies now. Ignoring this challenge may bring serious hurdles for public policy and for governance. It is also indicated that countries which delay attending to this problem may have to spend large sum of capital on retrofitting through the help of 'Big Data Analytics' from USA and despite that are likely to lose important information. If this capacity to manage Big Data is enhanced then many policies would become self-reflective, participative and relatively more inclusive since access and content simplicity which is the key to BD would enlighten citizens as well as governments. For example, the Human Resource programmes use BD to match positions to existing employees. One big problem in organizations is that the employees' profiles generally do not match the positions they get posted in due to their self-descriptions. HR departments scour through social media profiles, blogs and online conversations across the internet where talents and special skills are discovered for organizational requirements. BD helps to find out all details about the employee to post him/her where best suited. The Big Data expert from the IBM Company, Jeff Jones says, "You need to let data speak to you" and this is possible only when the unstructured data is converted into structured data.

A Social Scientist's Understanding of AI and BD

BD and AI are strangely connected as a storage tank and the water supply! BD in itself may not reach further than a satisfaction that information exists. However, this information is beyond human capacity to decipher and that is where machine learning through AI helps interpret data, its impact, location, scale and speed. This space and time data serves as a warning, as an instruction or as a plan of action for governing institutions. AI and BD are currently overlapping and interconnected. Their relationship is also improving cognitive initiatives of public institutions in DM.

AI is a form of ICT based computing machines where these machines become cognitive trendsetters to suggest measures and correct decisions. A smart phone with GPS and auto check or voice or face recognition systems is a basic AI based human cognition function. Traditional computing apps may also respond but through human based deciphering, interpreting and reacting which may be much longer and complex to be always correct. This may also miss out emergencies, sudden occurrences and urgencies to be encountered. For example a plan of action on a suspected flood may need a big human team of experts to draw a plan of action, evacuation and escape whereas an AI enabled machine may do it in seconds and may do it with the incorporation of human behaviour of support, object, capacity and reactions. BD which cannot react on its stored and classified data is also the basis and a reason of action for AI. BD is like an arm chair intellectual while AI is the alert army!

The origin of Big Data can be traced to the earlier analytical philosophers who discovered the mathematical logic in the way language is used. Ludwig Wittgenstein's *Tractatus Logico-Philosophicus* (1921) and Bertrand Russell's 'logical atomism' in his *Principia Mathematica* (1925–27, with A. N. Whitehead) inspired a debate on the fundamental building blocks of thought processes or an endorsement of analysis through which a given domain of enquiry can be defined and recast in a manner that remainder of the truths could be derived or accessed. Their logic of analysis suggested that the way human beings express themselves in their language propositions paves the way for understanding the world more logically. Even the fundamental truths of arithmetic are nothing more than relatively stable ways of playing a particular language game. Big Data is a form of a revolution within ICT which paves the way for many more ideas to flow in as society advances.

It is said that the lunch table conversations during the mid-1990s at the Silicon Graphics featured the Chief Scientist John Mashey quite prominently. Douglas Laney, a veteran data analyst at Gartner, declared John Mashey as the 'father of Big Data'.[1] However, the origin of the term is from scattered sources but as Victor Mayer-Schonberger and Kenneth Cukier (2013) simplify the debate by suggesting that the term has originated from the many debates on astronomy and genomics, sciences where data storage, correlation and retrieval leads to major breakthroughs in our understanding of the universe and well-being of people. It becomes fairly clear that Big Data originates out of the fundamental building blocks of language and culture which can be referred to as its genetics. The new digital forms of communication—Web sites, blog posts, tweets—are often very different from the traditional sources for the study of words, like books, news articles and academic journals.

How Would Governance Benefit from AI and Big Data?

Flood related twitter activity, WhatsApp messages, photographs and images shared provide a quicker understanding of the location, timing and impact of a disaster. BD may present a 350,000-tweets-per-minute data but to make it meaningful institutions will have to move to AI and machine learning. One would notice that smart phones gradually start producing data as per the user's liking when some names, numbers, songs, speeches and images appear on a priority choice. An open source software platform AIDR (Artificial Intelligence for Digital Response) is built to filter and classify billions of social media messages to enhance capacity to draw digital maps, programmes, evacuation routes and dashboard instructions and predictions during emergencies, disasters and catastrophes.

Sophistication in Decision Making Tools

In earlier times the decision making involved no process except the whims and fancies of the rulers. Later it evolved into some scientific principles which formed the inflexible parameter of good decision making. Contesting this approach Herbert Simon indicated a behavioural approach to decision making but warned that a halo of preconceived thoughts around decision makers led to bounded rationality. Big Data minimizes the fuzziness of all approaches and brings logic of science in data corroboration, correlation, forecasting and predictability in decision making. It also helped in making policies more inclusive and decision making increasingly holistic, interdisciplinary and sustainable. Besides these issues, BD is also needed for improved risk management in business and in governance. A case is mentioned below.

In 2009 the Flu virus was discovered in USA. All strains were collected from the Bird Flu, Swine Flu and H1N1 and their correlation was established with the 1918 Spanish Flu which infected half a billion and killed tens of millions. The information had to be relayed back to central organizations and tabulated. This was a big challenge as officials visited this information only once a week which was a fatal time span for communicable disease spread. At such a time Google through its in-house BD Analytics made 50 m. Common searches that Americans share online and compared with the Communicable Disease Report Data on the spread of seasonal flu between 2003–2008. This correlation established a staggering

450 m. Different mathematical models in order to test the search terms and finally helped in finding a solution. Without BD Analytics this was almost impossible or would have taken so long that the whole exercise would have become irrelevant.

Diagnostic Capability

Monitoring patient's history, well-being documents, nature of circulatory systems and frequency of infection can strengthen microscopic-long distance robotics which has enormous scope in telemedicine especially in the third world and in army locations. It has the ability to detect nascent heart attacks, early stages of cancer and also management of insulin levels.

AI and BD have contributed to the Food and Drug Administration of USA in many ways; i.e., Proteus Digital Health, a California based biomedical firm could kick-start the use of an electronic pill. It creates information which helps tissue engineering, genetic testing, DNA sequencing and source based solutions as well as early warning alerts on the basis of information corroboration and analytics.

Climate Change Related Early Warning Mechanism Systems

Climate change has brought substantial justification to have BD availability. The increasing inter-sectoral and inter-agency information such as the land, air and water bodies related changes, cloud formation, cyclones and hurricanes centred specialized data for over many hundred years and relationships to aquifers, flora and fauna, disasters and droughts, weather and crops etc. This expanse of information and the widening scope of its applicability in public policy have never existed prior to BD. Currently there are data and also the country and region based information which is scattered and much less accessed even during the period when the problem actually strikes. The meteorological data, density of population inhabitations, ecosystem services, local responses in the past to similar issues and urban planning records would combine in BD analytics to justify and enable retrofitting in decision making during troubled times of climate change.

Indispensability of 'Big Data' for Public Institutions

For many reasons, Big Data is becoming an unavoidable fact of governance in present times. Governance being an overlapping team work between public, private and non-state philanthropic enterprises, organizations need to find better ways to tap into the wealth of information hidden in this explosion of data around them to improve their competitiveness, efficiency, insight, profitability and more (Eaton, Deroos, Deutsch, Lapis, & Zikopoulos, 2012). The realm of BD as Eaton and his group of IBM experts suggest is the analysis of all data (structured, semi-structured and un-structured) so that quick access to relevant information becomes easier for everyone. As Big Data experts have revolved around many 'Vs', it would be interesting to look into some of them here.

The volume of data being created every day is breaking through the storage spaces. In 2003 it was 14 trillion in a day which required five exabytes of space. This volume was produced in two hours in 2011 and 10 minutes in 2013. For an average service to 100 million customers, Customer Service Providers would need 50 terabytes of location data daily. If stored for 100 days it would need five petabytes as almost five billion records are created in a day. In 2010 in US records, the most popular service provider company AT&T had 193 trillion Customer Data Records (CDR) in its database. The velocity of the data is also increasing. The global mobility data is growing at 78 percent of a compounded growth rate. Cisco Visual Networking Index (VNI-2013-2018), an ongoing initiative to track and forecast the impact of visual networking applications found that, 'Traffic' from wireless and mobile devices will exceed traffic from wired devices by 2016. By 2016, wired devices will account for 46 percent of IP traffic, while Wi-Fi and mobile devices will account for 54 percent of IP traffic.[2] In 2013, wired devices accounted for the majority of IP traffic at 56 percent. Overwhelmingly, the Global Internet traffic in 2018 will be equivalent to 64 times the volume of the entire global Internet in 2005 which suggests that bureaucracy and public officials may have to revise and reframe their capacity which would not be limited by their non-availability in office or by their multifarious tours as excuses for not attending and responding to important queries. To understand that much of the global Internet traffic which would reach 14 gigabytes (GB) per capita by 2018, rising by 5 GB per capita in 2013[3] would require additional capacities in the offices of public officials including the ability for BD analytics. As analytics is increasingly being embedded in business processes by using data-in-motion with reduced latency yet the real time data[4] which has to be catered to immediately and with urgency in every government, e.g., www.turn.com capacity of 10 m/sec.

The variety of data is rising very fast in equivalence to its volume and velocity. The old time Data Warehouse Technology[5] used in the 1990s cannot be relevant anymore for the fact that public policy cannot depend upon an individual's understanding anymore. Besides a technically efficient administrator, what is also be needed is an equivalent expansion of key government offices towards an adoption of latest reporting tools, data mining tools (SPSS, etc.) and GIS to name a few. The data would come from various sources and would be transformed using Extract Transform Load[6] (ET) data inside the Warehouse. In earlier times this could be possible by untrained or less trained 'babudom' as it was more or less a structured content but to allow the earlier capacity to continue would be to play havoc with public policy. The public policy spaces would then be littered with consultants, each one asking for their fee and pulling information to their vested commercial interests. Currently, data content is unstructured for lack of a directed objective. Once policy formulation begins differentiation within larger objectives; i.e., climate change as a main theme may add ever growing specificities such as coastal regulations, disaster risk reduction, ecosystem studies, disease control, food security and environmental changes then the need for Big Data to improve public policy formulation and implementation becomes important. To organize unstructured texts, sounds, social media blogs etc. government needs more enabling technology like the ones at IBMs Info-sphere stream platform.

Lastly but the most important requirement is the veracity (authenticity) of data for BD. Unlike governed internet data, BD comes from outside-our-control sources. Thus BD requires significant correctness and accuracy problems besides establishing and ensuring the credibility of data for target audience. Thus each Ministry of Government will have to first start with a basic data which routinely arrives at its posts and through analytics store it as Big Data. Right now much of the available data disappears or gets contaminated. Kevin Normandeau (2013) explains that BD veracity refers to the biases, noise and abnormality, the knowledge about which helps to clean the system. Many experts have added validity and volatility as important 'Vs' for BD. This may become important for the coming times when stored data could become outdated or irrelevant thereby suggesting a time period about its validity and also volatility. This is not so important for countries of South Asia which have yet to take their initial test drive on the BD highway.

Drivers for AI and BD Based Decision Making

In a compelling book of David Feinleib (2013) the author has tried to demystify Big Data as he emphasizes that to understand BD is to capture one of the most important trends of the present day world which surpasses every institutional boundary. The Changing governance paradigmatic requirements, e-governance expansion and rising number of internet and mobile users is a yeoman's task for routine administration to attend to. The new age citizen-customers are more sophisticated consumers who prefer to go online before taking a decision. Automation and convergence technology is speeding up faster with IVR, Kiosks and mobile telephony usages penetrating the regions untouched so far with any market or governance activity. Information is being collected through a hub-and-spoke model in a number of South Asian countries but BD and then creating its link to AI is still a distant priority in disaster management institutions. The base of AI and BD is the internet base in a country which in India is still weak and stands around 31 percent only till 2016 (IAMAI & KANTAR IMRB, 2016). Urban India has a coverage of 60 percent while the rural India the coverage is pathetically only 17 percent. Technology, trust and training are three 'Ts' which weaken adoption of new technologies which are AI and BD base.

Conclusion

An average annual multi-hazard risk loss in India alone is USD 88 billion out of a total average GDP of around USD 2690 billion (in 2018). South Asia has the world's largest number of poor and vulnerable community to be affected by recurrent disasters. This also obstructs and delays progress, sustainability and well-being of people. With this high rate of losses and damages an achievement of Sustainable Development Goals by 2030 may become impossible. Considering the region's voluminous governance challenges in terms of providing health care, livelihood, education, skills which is a basic platform to raise structures of disaster mitigation and risk reduction infrastructure, BD and AI can transform the sad scenario if management is well planned with experts and social scientists. There are many policy changes which have to be brought in through innovation, training and technology. Big Data is a mine of information to overcome and also escape many decisional catastrophes which are likely to come on the overloaded highway of government policies. This also requires balancing of a robust and secure public sector architecture that can accommodate the need for sharing data openly with all stakeholders in a transparent manner. This further entails a commitment from national governments to undertake a sincere and serious minded leadership in disaster management in the direction of new technology adoption which not only shares the human load of decision making but also brings greater accountability, transparency and cost-effective disaster management.

Notes

1. Lohr, Steve (2013) The Origins of 'Big Data': An Etymological Detective Story, New York Times, Feb. 1. Accessed http://bits.blogs.nytimes.com/2013/02/01/the-origins-of-big-data-an-etymological-detective-story/, 15.7.2014.
2. VNI Report available at http://www.cisco.com/c/en/us/solutions/collateral/service-provider/ip-ngn-ip-next-generation-network/white_paper_c11-481360.html
3. http://www.cisco.com/c/en/us/solutions/collateral/service-provider/ip-ngn-ip-next-generation-network/white_paper_c11-481360.html
4. Real-time data denotes information that is delivered immediately after collection. There is no delay in the timeliness of the information provided. It is of immense use to public officials as the 'Real-time data' is often used for navigation or tracking.
5. A data warehouse is the data repository of an enterprise. It is generally used for research and decision support. For further details see Joseph M. Wilson's 'An Introduction to Data Warehousing'(a PPT from Storet Co.) and Samii, Massood (2004) International Business and Information Technology: Interaction and Transformation in the Global Economy, New Hampshire USA: Psychology Press.
6. ETL suggests three functions; extract, transform, load, combined together into one tool to pull data out of one database and transfer it to another database. Extract is the process of reading data from a database. Transform is the process of converting the extracted data from its previous form into the form it needs to be in so that it can be placed into another database. Transformation occurs by using rules or lookup tables or by combining the data with other data. Load is the process of writing the data into the target database. This helps to either to shift data to data warehouse or to convert it into data marts which would store data for future usage as well as for marketing.

References

Eaton, C., Deroos, D., Deutsch, T., Lapis, G., & Zikopoulos, P. (2012). *Understanding Big Data, Analytics for Enterprise Class Hadoop and Streamlining Data*. New York: McGraw Hill.

Feinleib, D. (2013). *Big Data Demystified: How Big Data Is Changing the Way We Live, Love and Learn*. San Francisco: Big Data Group LLC.

IAMAI & KANTAR IMRB. (2016). *Internet IAMAI in India -IMRB Report*. Retrieved from http://bestmediainfo.com/wp-content/uploads/2017/03/Internet-in-India-2016.pdf

Mayer-Schonberger, V., & Cukier, K. (2013). *Big Data: A Revolution that Will Transform How We Live, Work and Think*. New York: Houghton Mifflin Harcourt Publishing Co.

Normandeau, K. (2013, September 12). *Beyond Volume, Variety and Velocity Is the Issue of Big Data Veracity*. Inside Big Data. Available at http://inside-bigdata.com/2013/09/12/beyond-volume-variety-velocity-issue-big-data-veracity/. Accessed 20 June 2014.

Schmidt, E., & Cohen, J. (2013). *The New Digital Age: Reshaping the Future of People, Nations and Business*. New York: Alfred A. Knopf, Random House Publication.

PART I

New Technologies in Disaster Management

CHAPTER 2

Artificial Intelligence and Early Warning Systems

Rabindra Lamsal and T. V. Vijay Kumar

Introduction

Disaster is a severe disruption or deviation from the norm, occurring usually for a short period, impacting the community and society as a whole, while causing widespread harm or damage to human, wildlife, environment, infrastructure and economy. Disaster management is a multi-faceted process to mitigate, respond to and recover from the consequences of a disaster. A disaster occurs when a hazard affects a population of vulnerable people. In other words, a disaster is a result of a combination of vulnerability, hazard and the inability to cope up with its negative consequences (IFRC, 2018). Researchers working in the Disaster Risk Reduction (*DRR*) domain share a common perception about all disasters, i.e. that they are generally man-made and that proactive human actions taken

R. Lamsal (✉) • T. V. V. Kumar
School of Computer and Systems Sciences, Jawaharlal Nehru University, New Delhi, India

Special Centre for Disaster Research, Jawaharlal Nehru University, New Delhi, India

T. V. V. Kumar, K. Sud (eds.), *AI and Robotics in Disaster Studies*, Disaster Research and Management Series on the Global South,
https://doi.org/10.1007/978-981-15-4291-6_2

before the occurrence of a hazard can prevent it from turning into a disaster. Therefore, almost all the disasters are attributable to human failure in defining and implementing emergency management measures (Blaikie, Cannon, Davis, & Wisner, 2005).

A hazard may also result in a secondary hazard that has greater impact; such as an underwater earthquake causing a Tsunami, which results in coastal flooding (inundation of coastal areas). Hazards are mainly categorized as natural or man-made. A natural hazard is a natural process or phenomenon, which includes events such as earthquakes, landslides, floods, blizzards, tsunamis, cyclones, hurricanes, heat waves, lightning strikes, tornadoes, volcanic eruptions etc. These hazards have a high probability of turning into disasters, as they may claim thousands of lives and may result in the destruction of environmental habitats and property. Man-made hazards are the consequences of technological or human actions. Such hazards include bioterrorism, fires (urban), explosions, collisions, structural collapses, wars, nuclear radiation leakages etc. Table 2.1 shows the classification of various hazards.

Table 2.1 Classification of hazards (World Health Organization International, 2002)

Natural hazards	Sudden occurrence (Monocausal)	Storm Heat wave Freeze Earthquake Volcanic eruption
	Progressive occurrence (multicausal)	Landslide Drought Flood Epidemic Pest
Man-made hazards	Sudden occurrence (Monocausal)	Fire Explosion Collision Shipwreck Structural collapse Environmental pollution
	Progressive occurrence (multicausal)	War Economic crisis

Disaster Management

Disaster Management is an integrated process of planning, organizing, coordinating and implementing necessary measures to perform the following (The Disaster Management Act, 2005):

- Preventing the threat of any potential hazard
- Mitigating the risks of any hazard and its consequences
- Building capacity
- Preparing for dealing with any hazard
- Giving prompt attention in response to any threats relating to a hazard
- Assessing severity of the effects of any hazard; evacuation; rescue and relief
- Rehabilitating and reconstructing

In simple terms, Disaster Management can be understood as the process of organizing and managing resources and responsibilities related to humanitarian aspects during the occurrence of hazards in order to minimize the negative impact of the disaster. Disaster management involves various organizational and operational decisions including prevention, mitigation, preparedness, response, recovery and rehabilitation.

The various key phases of activity within disaster management are:

1. **Pre-Disaster**: The pre-disaster phase includes activities meant to reduce the potential loss of lives and minimize the impact of the disaster on the habitat and property is likely to be impacted to minimize the losses when a hazard occurs. Activities relating to prevention, mitigation and preparedness fall inside this phase.
2. **During-Disaster**: The during-disaster phase includes activities carried out to meet the needs and provisions of the affected population and to alleviate and minimize the distress. Response-related activities fall inside this phase.
3. **Post-Disaster**: The post-disaster phase involves activities performed to achieve rapid and durable recovery, and incorporates emergency relief, rehabilitation of the affected community and the reconstruction of the destroyed property.

The major goals of disaster management include the minimization of losses from hazards, by ensuring a prompt response to the affected

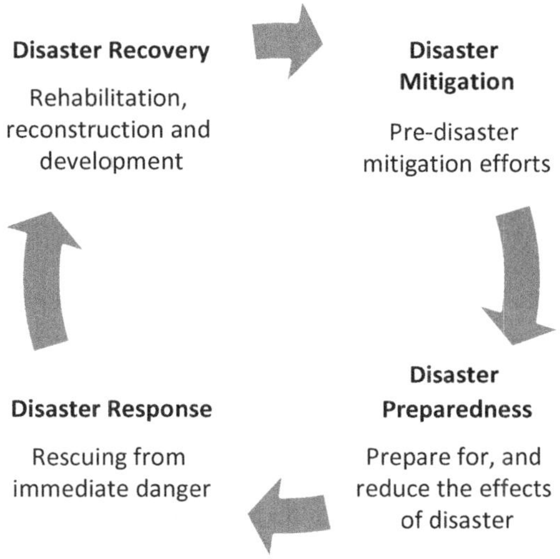

Fig. 2.1 The disaster management cycle

population and achieving optimum recovery. The disaster management cycle, depicted in Fig. 2.1, considers these goals and illustrates an ongoing process by which decision makers can plan for, and take effective decisions, to reduce the impact of the disaster and to appropriately act during an ongoing disaster and recover from the after-effects of the disaster. Effective decisions taken during each phase in the disaster management cycle ensures better preparedness, provides early warnings, reduces vulnerability and prevents the extent of disaster in the subsequent iteration of the cycle. This paper focuses on early warnings about the hazards that need to be disseminated to the vulnerable communities in time in order to prevent/ minimize the loss or damage due to potential disasters. This paper discusses early warning systems and the role played by artificial intelligence to design such systems.

Organization of the Paper

The organization of the paper is as follows. Section "Early Warning Systems" briefly discusses Early Warning Systems. AI techniques that can be used to design such systems are discussed in section "Artificial Intelligence" followed by the existing AI based early warning systems in section "*AI* and *EWS*". Section "Conclusion" is the conclusion.

Early Warning Systems

Early warning systems (EWSs) are designed to disseminate relevant information effectively and efficiently, as alarms or warnings, to communities at risk during or before a disaster event so that adequate steps can be taken to minimize the loss and damage associated with disasters. The four interacting elements, which need to be incorporated inside an *EWS* for it to be effective and complete, are risk knowledge; monitoring and warning service; dissemination and communication; response capability (Basher, 2006). Amongst these, the second element, viz. monitoring and warning service, is a key component of *EWS*. However, accurate predictions may not be enough to reduce the impact of hazards. The effectiveness of *EWS* depends on strong political commitment, robust institutional capacities and public awareness (Basher, 2006). The early warnings generated by an *EWS* must reach the vulnerable communities well in time, as an effective step towards disaster preparedness. The human factor plays a key role in maintaining the sustainability of *EWS* (Twigg, 2003).

Design and development of *EWS* require a broad multidisciplinary approach with a knowledge base of technology, geoscience and social science (Basher, 2006). Technology incorporates the use of *AI* techniques for developing early warning models or some intelligent knowledge-based systems. Geoscience deals with the study of Earth, its oceans, atmosphere, rivers and lakes; and in current times Geographic Information System (*GIS*) and Remote sensing have shown significant improvements in monitoring and analyzing the physical constitution of the Earth and its atmosphere. Although advancements in technology have led to quicker predictions of hazards, the disaster preparedness phase of the disaster management cycle requires an understanding of human behavior particularly before or during the occurrence of a disaster, which has a social science perspective. The various aspects, which need to be considered in the design and development of an *EWS*, incorporating the three disciplines mentioned above include (Basher, 2006):

- Use of geospatial models and risk maps
- Cost effective monitoring systems
- Data assimilations from sensors
- Improvements in the accuracy of prediction models
- Development of warning decision tools for administrators
- Evaluation of warning communication methods

- Study of human response behavior
- Visualization of impacts caused by a hazard

This paper focuses majorly on the technological aspect of *EWSs* with the focus on the use of *AI* for developing such systems. The wide use of *AI* having a significant impact on society and is being seen to be playing a major role in mitigating and preparing for disasters. The applications of *AI* can be used effectively in developing EWSs for disaster mitigation and preparedness. AI techniques that can be used for designing and developing *EWSs* are discussed next.

Artificial Intelligence

Artificial intelligence (*AI*) is the intelligence exhibited by computer systems in contrast to the intelligence displayed by humans (Russell & Norvig, 2016). In other words, *AI* is the simulation of human intelligence processes by computer systems (Rouse, 2010). The *AI* process includes knowledge representation, planning, learning (the acquisition of information and rules for using the information), reasoning (applying the rules to point towards a conclusion), self-correction, natural language processing, and the ability to move and manipulate objects. The high-profile applications of *AI* include speech recognition, autonomous vehicles, verification of mathematical theorems, medical diagnosis, personal voice assistants, search engines, spam filtering, predictive modeling and online targeted advertisements (Lohr, 2016; Russell & Norvig, 2016; The Economist, 2016). Figure 2.2 lists the various subfields of AI.

There are, in general, two types of *AI*: weak *AI* and Strong *AI*. Weak *AI* uses the principles of simulating human intelligence to carry out one

Machine Learning	Natural Language Processing	Speech
Supervised Unsupervised Deep Learning	Translation Classification & Clustering Information Extraction	Text to Speech Speech to Text
Expert Systems	**Vision**	**Others**
Inference Engine Knowledge Base	Image Recognition Machine Vision	Planning Robotics

Fig. 2.2 Subfields of Artificial Intelligence

specific task. Whereas Strong *AI* seeks to imitate human intelligence for completing any task, just like a human. Strong *AI* can find a solution to any unfamiliar problem without human intervention. The advancements in the field of *AI* have resulted in the evolution of several research areas. The various subfields of *AI* include machine learning, natural language processing, speech recognition, expert systems, vision, planning and robotics of which machine learning techniques have been widely used for designing Early Warning Systems. Machine learning techniques that can be used for designing and developing *EWSs* are discussed next.

Machine Learning

Machine learning is a subfield of Artificial Intelligence, which provides computer systems with an ability to train and improve themselves without explicitly being programmed. The learning is based on patterns and inferences. Machine learning algorithms formulate a mathematical model using a training data set and make predictions on unseen data using the developed model. Machine learning algorithms are majorly categorized into supervised and unsupervised learning. Supervised learning algorithms use labeled data to predict unseen data whereas, unsupervised learning algorithms work on unlabeled data and are used for inferring the hidden structure in the dataset. Supervised learning includes prediction tasks, such as regression (when the output variable is a real-continuous value) and classification (when the output variable is a category or a class), while unsupervised learning includes tasks, such as clustering and association rule mining. To distinguish between *AI* and machine learning, if a written computer program exhibits human-like behavior, it can be *AI*. It is called machine learning only when the essential parameters are learnt from the data automatically (Genzel, 2016). Some of the widely used machine learning techniques that can be used for designing and developing EWSs are briefly discussed below:

Regression Regression is a supervised machine learning technique of approximating a mapping function f(x) from the independent variable (X) to a real-continuous dependent variable (y). For a given set of n independent variables $x_1, x_2, x_3, \ldots, x_n$, a dependent variable Z, and the parameters $b_0, b_1, b_2, \ldots, b_n$; a multivariate linear regression model takes the form:

$$Z = b_0 + b_1 * x_1 + \ldots + b_n * x_n$$

Since the regression model predicts a continuous value, the robustness of the model is evaluated, based on the difference between the actual output and the predicted output. The cost function of a linear regression model is given by:

$$cost\ function = \frac{1}{n}\sum_{i=1}^{n}\left(y_{(i)} - Z_{(i)}\right)^2$$

To reduce the cost function value, *Gradient Descent* method is used (Ruder, 2016). *Gradient Descent* method is an optimization algorithm, used during training of a machine learning model, for readjusting the parameters iteratively that drives the cost function to its global minima (Ruder, 2016). For a given cost function, a random starting point is selected and the slope (gradient) is computed at that point with respect to each parameter that influences the cost function. Based on the slope (negative or positive), each parameter is updated accordingly and this process continues until the global minima is reached.

Classification Classification is a supervised machine learning technique (Bishop, 2006; Sebastiani, 2002) of determining the most appropriate category of a dependent variable, based on one or more independent variables. In other words, classification is an act of predicting a discrete class. The learning algorithm learns from the past data and uses this learning to classify new data to a particular class/category/discrete response. The widely used classification-based traditional machine learning algorithms include Logistic Regression, Naïve Bayes, Decision Trees, Random Forests, k-Nearest Neighbor, Artificial Neural Network (Kotsiantis, Zaharakis, & Pintelas, 2007; Witten, Frank, Hall, & Pal, 2016). These algorithms are discussed below:

Logistic Regression Logistic Regression (Cox, 1958; Walker & Duncan, 1967) is one of the commonly used machine learning algorithms for classification tasks. For a given set of n independent variables $x_1, x_2, x_3, \ldots, x_n$, a dependent variable Z, and the parameters $b_0, b_1, b_2, \ldots, b_n$; the logistic model takes the form:

$$Z = b_0 + b_1 * x_1 + \ldots + b_n * x_n$$

Unlike in Linear Regression, the output of the Logistic Regression is strictly between 0 and 1. Hence, the value of *Z* is given as input to the *Sigmoid* function.

$$h_\theta(x) = sigmoid(Z)$$

$$h_\theta(x) = \frac{1}{1 + e^{-(b0 + b1*x1 + \ldots + bn*xn.)}}$$

Where, $h_\theta(x)$ is the hypothesis of Logistic Regression.

The cost function used in Linear Regression cannot be used for Logistic Regression, as it might end up being a non-convex function with many local minima and it might become difficult to reach the global minimum. Hence, for Logistic Regression, the cost function is defined as:

$$cost(h_\theta(x), y) = \begin{cases} -\log(h_\theta(x))\, if\ y = 1 \\ -\log(1 - h_\theta(x))\, if\ y = 0 \end{cases}$$

The above two functions can be represented as a single function:

$$J(\theta) = -\frac{1}{m}\Sigma\left[y^{(i)}\log(h_\theta(x(i))) + (1 - y^{(i)})\log(1 - h_\theta(x(i)))\right]$$

Gradient Descent method is used to minimize the cost function.

$$\theta_j := \theta_j - alpha * \frac{\partial j(\theta)}{\partial x}$$

Where, *alpha* is the *learning rate*.

Naïve Bayes Naïve Bayes is a conditional probabilistic classifier (Langley, Iba, & Thompson, 1992), based on Bayes Theorem, which assumes that the features are conditionally independent. Naïve Bayes classifiers are extremely fast when compared with more sophisticated classifiers like Support Vector Machines (*SVM*). Given an instance with a set of *n* independent variables x_1, x_2, x_3, …, x_n, the model assigns probabilities to the instance for every class.

$$P(y|x_1, x_2, x_3, \supset, x_n)$$

This conditional probability can be decomposed using the Bayes' Theorem, as given below:

$$P(y|X) = \frac{P(X|y)P(y)}{P(X)}$$

Where, $P(y|X)$ is the *Posterior Probability*, $P(X|y)$ is the *Likelihood*, $P(y)$ is the *Class Prior Probability* and $P(X)$ is the *Predictor Prior Probability*. $X = (x_1, x_2, x_3, x_4, \ldots, x_n)$ represents the parameters or features. By substituting for X and expanding using the chain rule, the Posterior Probability becomes:

$$P(y|x_1, \ldots, x_n) = \frac{P(x_1|y)P(x_2|y)\ldots P(x_n|y)P(y)}{P(x_1)P(x_2)\ldots P(x_n)}$$

Since the denominator does not change for all entries in the dataset, it can be removed and proportionality can be introduced.

$$P(y|x_1, \ldots, x_n) \propto P(y)\prod_{i=1}^{n} P(x_i|y)$$

In the multi-class classification problem, the class y with the maximum probability is identified using the function given below.

$$y = argmax_y P(y)\prod_{i=1}^{n} P(x_i|y)$$

Random Forests Random Decision Forests (Ho, 1995) is an ensemble learning method, which is used for both regression and classification purposes. This method incorporates the approach of constructing multiple numbers of decision trees using the random subset of the training dataset and assigning the new data instances to the class, having the majority of the votes. The generation of multiple trees ensures the reduction of high bias, i.e. overfitting. The Random Forests can handle large datasets of high dimensions and outperform decision trees in most of the cases, since a single decision tree is more likely to be affected by the presence of noise in the dataset.

Random Forests use the *Bagging* approach to randomly create random *n* subsets of a given dataset. Around *one-third* of the samples of the original dataset are left out, known as *Out-of-bag* samples. Multiple decision trees are constructed using the *n* subsets of the original dataset. Each tree is generated based on a different subset. The final prediction is made, based on *averaging* for regression problems and *voting* for classification problems.

k-Nearest Neighbor (kNN) *k*-Nearest Neighbor (Cover & Hart, 1967) is one of the simplest classification-based learning algorithms, which classifies new instances based on a similarity measure (distance functions). Euclidean Distance, Manhattan Distance and Minkowski Distance (Hu, Huang, Ke, & Tsai, 2016) are the most commonly used distance functions in case of continuous variables. While Hamming Distance is used for categorical variables. A new instance is classified by a majority vote of its neighbors, with the instance being classified to the most common class among its *k* nearest neighbors, based on the computed distance. The best value of *k* is identified experimentally.

The Euclidean Distance between points *p* and *q* for an *n*-dimensional space is given by:

$$d(p,q) = \sqrt{\sum_{i=1}^{n} (p_i - q_i)^2}$$

While the Manhattan Distance between points *p* and *q* for an *n*-dimensional space is computed using the following formula:

$$d(p,q) = p - q_1 = \sum_{i=1}^{n} |p_i - q_i|$$

The working of *kNN* is as follows. The distance between a new instance and each training example is computed. The computed distances are sorted in ascending order and the number of *k* neighbors to be considered is determined. Based on the number of *k* neighbors, the top *k* rows are selected from the sorted list. The most frequent class amongst the rows in the list is considered as the target class for the new instance that is to be classified.

Artificial Neural Network (ANN) Artificial Neural Network is a machine learning method (Haykin, 1994), which is modeled roughly after the biological neural networks. *ANN* is used for both classification and clustering purposes. *ANN* is a network of artificial neurons, which are connected by synapses, i.e. the weights. Each artificial neuron is a computational unit. Each neuron in the network receives inputs and computes the activation function, based on the inputs, to determine its state, and to finally compute the output using the inputs and its activation value.

ANN comprises *n* layer(s) of connected computational units called neurons. A trained network takes a new instance, applies the activation function to the weighted aggregation of the instance for each neuron, in every hidden layer, and then outputs the prediction for the given instance. The first layer of the network is the input layer, which takes the inputs and forwards the weighted aggregation of the inputs, as inputs, to the next layer (hidden layer) in the network. In the hidden layer, an activation function is applied to the weighted aggregation of inputs to compute the activation value, which becomes the input to the next hidden layer. This process continues until the output layer, i.e. the last layer in the network, processes the aggregated inputs from the previous layer through an activation function to form the output of the neural network.

Deep Learning It is a subfield of machine learning, based on artificial neural networks. Deep learning methods are representation-learning methods, which include multiple depths of representation (*n* number of hidden layers) composed of simple but non-linear modules, where each module transforms the representation at one level into a representation at higher level (LeCun, Bengio, & Hinton, 2015). Unlike other machine learning methods, deep learning methods enable machines to discover complex structures in the raw dataset automatically. The learning can be supervised, unsupervised and semi-supervised. Deep learning is making significant advancements in fields such as computer vision, speech recognition, natural language processing, machine translation, medical image analysis.

AI has been effectively used in designing and developing *EWSs*. An overview of some well-known *AI* based *EWSs* is discussed next.

AI AND *EWS*

AI and its applications provide consummate solutions for design and development of Early Warning Systems. *AI* can also be applied effectively in Geoscience. *GIS* and Remote Sensing have been used for monitoring landscapes (Liu & Yang, 2015; Vogelmann et al., 2010) and coastlines (Gens, 2010). Trained *AI* models can learn from, and analyze, the data available from *GIS* and Remote Sensing systems to build intelligent knowledge-based systems or some ideal expert system for identifying any abnormal behavior in order to predict the possibility of occurrence of disasters. Satellite imagery has also been used in the prediction of hazards (Bischke et al., 2017; Carrara, Guzzetti, Cardinali, & Reichenbach, 1999; Gillespie, Chu, Frankenberg, & Thomas, 2007; Kung, Chen, & Ku, 2012; Nayak & Zlatanova, 2008; Tadesse, Brown, & Hayes, 2005) and infectious disease outbreaks (Ford et al., 2009). The quantitative data obtained from the sensors and seismographs helps in designing complex learning models to predict the occurrence of possible disaster events such as Tsunamis, Earthquakes and Volcanic eruptions, so that the decision makers can take proactive measures like evacuation of the people living in the vulnerable areas.

Wireless Sensor Networks (*WSNs*) have a great potential for infrastructure monitoring applications (Cheekiralla, 2005). The sensors attached to infrastructures, such as large foundations, roads, bridges, tunnels and dams can detect even a millimeter scale of deformation. The data received from such sensor networks can be monitored in real-time to avoid disasters, which might occur due to the collapse of buildings, tunnels or bridges.

Social media has been seen as an excellent communication platform for dissemination of warnings for any possible disaster event (Chatfield & Brajawidagda, 2012). A governmental agency *BMKG*, Indonesia (Badan Meteorologi, Klimatologi, dan Geofisika), generated a tweet related to the possibility of a Tsunami event just six minutes and seven seconds after a powerful earthquake occurred. Interestingly, the reach of tweet went beyond 4 million within fifteen minutes of the occurrence of the earthquake. This shows that social media platforms, such as Facebook and Twitter, having a userbase of millions can play an important role in the dissemination of timely warnings to the communities at risk.

The German-Indonesian Tsunami Early Warning System is a high-profile example of implemented *EWS* for the Indian Ocean, which uses seismometers, *GPS*, tide gauges, ocean-bottom pressure sensors and *GPS* Intelligent Buoys for functioning (Rudloff, Lauterjung, Münch, & Tinti, 2009). The developed *EWS* is capable of determining the location, the magnitude and the depth of an earthquake and disseminate the related information to vulnerable communities as quickly as possible within five minutes of its occurrence. The system aims at delivering early warnings to the vulnerable communities, as quickly as possible. In a similar manner, a Tsunami *EWS* has been designed to make a decision related to warnings or evacuations, based on the data obtained from the sensors deployed in the deep ocean (Liu, Wang, & Salisbury, 2009). For this purpose, the region of interest is divided into multiple segments and the data is collected, processed and analyzed. Based on this, a decision is taken whether to issue warnings and/or to perform evacuations of the vulnerable communities.

Researchers from Google and Harvard University have used an enormous amount of seismic data to train a neural network model for predicting the aftershocks of earthquakes (DeVries, Viégas, Wattenberg, & Meade, 2018). The model was trained to analyze and understand the patterns of mainshock–aftershock events on a dataset of more than 1,31,000 such events. When tested on 30,000 mainshock–aftershock pairs, the trained neural network model predicted the aftershocks for the independent test-set more accurately in comparison to the existing methods.

Google has been using *AI* for developing forecasting models that predict when and where floods might occur (Matias, 2018). The models are trained using historical data, river level data, terrain and elevation data. The developed flood forecasting models generate maps and run thousands of simulations in each possible location to predict the time, the place and the severity of the flooding event. Similarly, Microsoft and *IBM* are also working on developing *AI* models for predicting the occurrence of potential natural disasters (Gourgey, 2018; PTI, 2018).

A couple of research works have also been carried out in the Indian coastal zones. In Ali, Kishtawal and Jain (2007), an *ANN* model was trained using twelve hours of observations to predict the possible path of a cyclone over the Indian Ocean, 24 hours in advance. Similarly, Sahoo and Bhaskaran (2019) trained an *ANN* for predicting a storm surge and onshore flooding. The developed model was validated over the historical storm-tide and onshore flooding instances of 1999 Odisha Super Cyclone. Further, Poompavai and Ramalingam (2013) carried out the geospatial analysis for coastal risk assessment for cyclones. The satellite and GIS data are used to create thematic layers and hazard maps. Coastal risk indices are generated for each component of risk using the analytic hierarchy process. An efficient warning system can be designed based on the risk index of a region and the pattern predicted for a cyclone. Identification of evacuation roots was a focus area of such *EWS*.

*WSN*s have also become a part of technologies for designing *EWS*s. *WSN*s consist of small and low-cost sensor nodes distributed over an area of interest. The nodes are capable of sensing, processing, communicating and exchanging sensory data within the network. This capability of *WSN*s includes the functionality of event detection, which can be of great help in the detection of near real-time events, such as meteorological disasters. Bahrepour, Meratnia, Poel, Taghikhaki and Havinga (2010) trained decision tree classifiers for early detection of residential fires using *WSN*s. Each sensor node in the network comprises a decision tree classifier, and a voting approach was used to reach a decision based on the decision made by each sensor node at the individual level. Similarly, *UrbanFlood* (Pyayt et al., 2011) also uses the sensors within flood embankments for providing online *EWS* and real-time emergency management.

Ventilating systems are an essential part of underground mining systems (Cheng, Yang, & Luo, 2010). The ventilating systems provide a sufficient amount of fresh air to dilute contaminants, such as methane, in order to maintain a suitable environment within the mining system in order to prevent possible disasters. Various works have been carried out in designing models for monitoring ventilation systems. Rough Set (*RS*) theory and machine learning algorithms, such as *ANN* and Support Vector Machines (*SVM*), have been used in designing early warning models for the mine ventilation system (Cheng & Yang, 2012; Zhao & Wang,

2009). Internet-of-Things (*IoT*) based early warning system has also been designed for providing safety and improving services in underground coal mines (Jo & Khan, 2017). Numerous sensor modules were installed at various places in underground mines to measure ***temperature***, ***humidity***, CO_2, CO and CH_4 for monitoring the safety status of the underground mine using "*Mine Warning Index*". Similarly, *WSNs* have been effective in monitoring and providing early warnings of a possible disaster, such as detecting structural variations caused by underground collapses, in underground coal mines (Li & Liu, 2009).

AI has also been used in detecting potential crowd disasters. Researchers from Baidu Big Data Lab (Zhou, Pei, & Wu, 2018) have proposed an *EWS* which is capable of detecting possible crowd disasters 1 to 3 hours in advance. Based on users' search query on the Baidu map and the number of positioning users in a particular area, the system invokes warnings for possible crowd specific disastrous events ahead of time. The system further uses a machine learning model to measure the risks associated with such events.

Overall, AI based EWSs have played a key role in disaster mitigation and preparedness. Also, with the advent of Internet of Things and advanced technology driven sensor devices, more and more data can be captured and analyzed, which in turn would enable the above EWSs to effectively and efficiently minimize the severe consequences of disasters.

Conclusion

The spread and use of *AI* are already having significant impact on society and can play a major role in the disaster preparedness phase of the disaster management cycle. The use of AI can be further enhanced to develop a scalable and real-time EWS by integrating imprecise and uncertain data to extract dynamic patterns, interpret and visualize them for time-sensitive decisions. AI have been used to develop EWS to address the following problems:

- Detecting location, magnitude and depth of an earthquake, as soon as it occurs
- Predicting aftershocks of earthquakes
- Forecasting when and where floods might occur
- Simulating the path of cyclones in advance
- Predicting storm surges and inundations

- Detecting near real-time events such as meteorological disasters using Wireless Sensor Networks (*WSNs*)
- Monitoring underground mine ventilation systems and detecting structural variations caused by underground collapses
- Detecting potential crowd related disasters such as stampede

With the use of *AI*, early warning systems can monitor and warn about a possible disastrous event, be it natural or man-made, to vulnerable communities and various stakeholders so that timely proactive and preventive actions can be taken to minimize the loss of lives and to limit the damage to property due to disasters.

References

Ali, M. M., Kishtawal, C. M., & Jain, S. (2007). Predicting Cyclone Tracks in the North Indian Ocean: An Artificial Neural Network Approach. *Geophysical Research Letters, 34*(4).

Bahrepour, M., Meratnia, N., Poel, M., Taghikhaki, Z., & Havinga, P. J. M. (2010). Distributed Event Detection in Wireless Sensor Networks for Disaster Management. *2010 International Conference on Intelligent Networking and Collaborative Systems*, 507–512.

Basher, R. (2006). Global Early Warning Systems for Natural Hazards: Systematic and People-Centred. *Philosophical Transactions of the Royal Society A: Mathematical, Physical and Engineering Sciences, 364*(1845), 2167–2182.

Bischke, B., Bhardwaj, P., Gautam, A., Helber, P., Borth, D., & Dengel, A. (2017). Detection of Flooding Events in Social Multimedia and Satellite Imagery Using Deep Neural Networks. *MediaEval.*

Bishop, C. M. (2006). *Pattern Recognition and Machine Learning.* New York: Springer.

Blaikie, P., Cannon, T., Davis, I., & Wisner, B. (2005). *At Risk: Natural Hazards, People's Vulnerability and Disasters.* Routledge.

Carrara, A., Guzzetti, F., Cardinali, M., & Reichenbach, P. (1999). Use of GIS Technology in the Prediction and Monitoring of Landslide Hazard. *Natural Hazards, 20*(2–3), 117–135.

Chatfield, A., & Brajawidagda, U. (2012). *Twitter Tsunami Early Warning Network: A Social Network Analysis of Twitter Information Flows.*

Cheekiralla, S. (2005). Wireless Sensor Network-Based Tunnel Monitoring. *Proceedings of the Real WSN Workshop.*

Cheng, J., & Yang, S. (2012). Data Mining Applications in Evaluating Mine Ventilation System. *Safety Science, 50*(4), 918–922.

Cheng, J., Yang, S., & Luo, Y. (2010). Mathematical Models for Optimizing and Evaluating Mine Ventilation System. *Proceedings of the 13th United States/ North American Mine Ventilation Symposium*, Sudbury, ON, Canada, 13–16.

Cover, T. M., & Hart, P. E. (1967). Nearest Neighbor Pattern Classification. *IEEE Transactions on Information Theory, 13*(1), 21–27.

Cox, D. R. (1958). The Regression Analysis of Binary Sequences. *Journal of the Royal Statistical Society: Series B (Methodological), 20*(2), 215–232.

DeVries, P. M. R., Viégas, F., Wattenberg, M., & Meade, B. J. (2018). Deep Learning of Aftershock Patterns Following Large Earthquakes. *Nature, 560*(7720), 632.

Ford, T. E., Colwell, R. R., Rose, J. B., Morse, S. S., Rogers, D. J., & Yates, T. L. (2009). Using Satellite Images of Environmental Changes to Predict Infectious Disease Outbreaks. *Emerging Infectious Diseases, 15*(9), 1341.

Gens, R. (2010). Remote Sensing of Coastlines: Detection, Extraction and Monitoring. *International Journal of Remote Sensing, 31*(7), 1819–1836.

Genzel, D. (2016). *What Is the Difference Between AI, Machine Learning, NLP, and Deep Learning?* Retrieved from Quora website: https://www.quora.com/ What-is-the-difference-between-AI-Machine-Learning-NLP-and-Deep-Learning/answer/Dmitriy-Genzel

Gillespie, T. W., Chu, J., Frankenberg, E., & Thomas, D. (2007). Assessment and Prediction of Natural Hazards from Satellite Imagery. *Progress in Physical Geography, 31*(5), 459–470.

Gourgey, B. (2018). How Artificial Intelligence Could Prevent Natural Disasters. *Wired*. Retrieved from https://www.wired.com/story/how-artificial-intelligence-could-prevent-natural-disasters/

Haykin, S. (1994). Neural Networks: A Comprehensive Foundation. Prentice Hall PTR.

Ho, T. K. (1995). Random Decision Forests. *Proceedings of 3rd International Conference on Document Analysis and Recognition, 1*, 278–282.

Hu, L.-Y., Huang, M.-W., Ke, S.-W., & Tsai, C.-F. (2016). The Distance Function Effect on K-Nearest Neighbor Classification for Medical Datasets. *Springerplus, 5*(1), 1304.

IFRC. (2018). *What Is a Disaster?* Retrieved from https://www.ifrc.org website: https://www.ifrc.org/en/what-we-do/disaster-management/about-disasters/what-is-a-disaster/

Jo, B., & Khan, R. (2017). An Event Reporting and Early-Warning Safety System Based on the Internet of Things for Underground Coal Mines: A Case Study. *Applied Sciences, 7*(9), 925.

Kotsiantis, S. B., Zaharakis, I., & Pintelas, P. (2007). Supervised Machine Learning: A Review of Classification Techniques. *Emerging Artificial Intelligence Applications in Computer Engineering, 160*, 3–24.

Kung, H.-Y., Chen, C.-H., & Ku, H.-H. (2012). Designing Intelligent Disaster Prediction Models and Systems for Debris-Flow Disasters in Taiwan. *Expert Systems with Applications, 39*(5), 5838–5856.

Langley, P., Iba, W. I., & Thompson, K. (1992). An Analysis of Bayesian Classiers. *Proceedings of the Tenth National Conference on Artificial Intelligence, 90*(415), 223–228.

LeCun, Y., Bengio, Y., & Hinton, G. (2015). Deep Learning. *Nature, 521*(7553), 436.

Li, M., & Liu, Y. (2009). Underground Coal Mine Monitoring with Wireless Sensor Networks. *ACM Transactions on Sensor Networks, 5*(2), 10:1–10:29. https://doi.org/10.1145/1498915.1498916

Liu, P. L.-F., Wang, X., & Salisbury, A. J. (2009). Tsunami Hazard and Early Warning System in South China Sea. *Journal of Asian Earth Sciences, 36*(1), 2–12.

Liu, T., & Yang, X. (2015). Monitoring Land Changes in an Urban Area Using Satellite Imagery, GIS and Landscape Metrics. *Applied Geography, 56*, 42–54.

Lohr, S. (2016, February). The Promise of Artificial Intelligence Unfolds in Small Steps. *The New York Times.* Retrieved from https://www.nytimes.com/2016/02/29/technology/the-promise-of-artificial-intelligence-unfolds-in-small-steps.html

Matias, Y. (2018). *Keeping People Safe with AI-Enabled Flood Forecasting.* Retrieved from Google Blog website: https://www.blog.google/products/search/helping-keep-people-safe-ai-enabled-flood-forecasting/

Nayak, S., & Zlatanova, S. (2008). *Remote Sensing and GIS Technologies for Monitoring and Prediction of Disasters.* Springer.

Poompavai, V., & Ramalingam, M. (2013). Geospatial Analysis for Coastal Risk Assessment to Cyclones. *Journal of the Indian Society of Remote Sensing, 41*(1), 157–176.

PTI. (2018). IBM to Invest in Tech to Predict Floods, Cyclones in India. *The Economic Times.* Retrieved from https://economictimes.indiatimes.com/tech/software/ibm-to-invest-in-tech-to-predict-floods-cyclones-in-india/articleshow/64319639.cms

Pyayt, A. L., Mokhov, I. I., Lang, B., Krzhizhanovskaya, V. V., Meijer, R. J., et al. (2011). Machine learning methods for environmental monitoring and flood protection. *World Academy of Science, Engineering and Technology, 78*, 118–123.

Rouse, M. (2010, November). AI (Artificial Intelligence). *Search EnterpriseAI.* Retrieved from https://searchenterpriseai.techtarget.com/definition/AI-Artificial-Intelligence

Ruder, S. (2016). An Overview of Gradient Descent Optimization Algorithms. *ArXiv Preprint ArXiv:1609.04747.*

Rudloff, A., Lauterjung, J., Münch, U., & Tinti, S. (2009). Preface "The GITEWS Project (German-Indonesian Tsunami Early Warning System)". *Natural Hazards and Earth System Sciences, 9*(4), 1381–1382.
Russell, S. J., & Norvig, P. (2016). *Artificial Intelligence: A Modern Approach.* Boston: Pearson Education Limited.
Sahoo, B., & Bhaskaran, P. K. (2019). Prediction of Storm Surge and Coastal Inundation Using Artificial Neural Network – A Case Study for 1999 Odisha Super Cyclone. *Weather and Climate Extremes, 23*, 100196.
Sebastiani, F. (2002). Machine Learning in Automated Text Categorization. *ACM Computing Surveys (CSUR), 34*(1), 1–47.
Tadesse, T., Brown, J. F., & Hayes, M. J. (2005). A New Approach for Predicting Drought-Related Vegetation Stress: Integrating Satellite, Climate, and Biophysical Data over the US Central Plains. *ISPRS Journal of Photogrammetry and Remote Sensing, 59*(4), 244–253.
The Disaster Management Act. (2005). Retrieved from https://www.ndmindia.nic.in/images/The Disaster Management Act, 2005.pdf
TheEconomist. (2016). Why Firms Are Piling into Artificial Intelligence. *The Economist Newspaper Limited.* Retrieved from https://www.economist.com/the-economist-explains/2016/03/31/why-firms-are-piling-into-artificial-intelligence
Twigg, J. (2003). The Human Factor in Early Warnings: Risk Perception and Appropriate Communications. In *Early Warning Systems for Natural Disaster Reduction* (pp. 19–26). Springer.
Vogelmann, J. E., Kost, J. R., Tolk, B., Howard, S., Short, K., Chen, X., et al. (2010). Monitoring Landscape Change for LANDFIRE Using Multi-Temporal Satellite Imagery and Ancillary Data. *IEEE Journal of Selected Topics in Applied Earth Observations and Remote Sensing, 4*(2), 252–264.
Walker, S. H., & Duncan, D. B. (1967). Estimation of the Probability of an Event as a Function of Several Independent Variables. *Biometrika, 54*(1–2), 167–179.
Witten, I. H., Frank, E., Hall, M. A., & Pal, C. J. (2016). *Data Mining: Practical Machine Learning Tools and Techniques.* Morgan Kaufmann.
World Health Organization International. (2002). *Disasters & Emergencies: Definitions.* Retrieved from http://apps.who.int/disasters/repo/7656.pdf
Zhao, Y., & Wang, H. (2009). Study on Ventilation System Reliability Early-Warning Based on RS-ANN. *2009 Sixth International Conference on Fuzzy Systems and Knowledge Discovery, 3*, 598–602.
Zhou, J., Pei, H., & Wu, H. (2018). Early Warning of Human Crowds Based on Query Data from Baidu Maps: Analysis Based on Shanghai Stampede. In Z. Shen & M. Li (Eds.), *Big Data Support of Urban Planning and Management: The Experience in China* (pp. 19–41). https://doi.org/10.1007/978-3-319-51929-6_2

CHAPTER 3

Artificial Intelligence in Disaster Management: Rescue Robotics, Aerial Mapping and Information Sourcing

Keshav Sud

Introduction to AI in Disaster Management

The general challenge that government officials and first-responders face in the aftermath of a disaster is the large-scale and time-sensitive rescue and recovery effort. For launching an effective effort the scale and type of damage need to be assessed, then the evacuation and emergency response is planned, areas for search-and-rescue missions are prioritized, and the kinds of medical aids needed are decided and mobilized. All of these tasks are extremely challenging as the affected and surrounding areas are generally difficult to access in the aftermath. This need to enable access to these areas while minimizing the monetary and personnel risk is fueling the development and adoption of rescue robotics and AI.

During the past decade, advancements in artificial intelligence (AI) have significantly extended our ability to predict disasters and provide assistance during disasters. AI-driven improvements can be seen in early

K. Sud (✉)
Data Science, Keurig Dr Pepper, Burlington, MA, USA

T. V. V. Kumar, K. Sud (eds.), *AI and Robotics in Disaster Studies*, Disaster Research and Management Series on the Global South,
https://doi.org/10.1007/978-981-15-4291-6_3

warning systems, crowdsourced information systems, rescue, and aid delivery. Although today AI has many faces, this chapter will be focusing on robotics and machine learning in the context of predicting disasters, and enabling faster and scalable rescue and aid delivery efforts.

The field of robotics and robots have existed over decades; however, the recent exponential growth in sensor and computation technology has led to the evolution of robots from low-level decision-making machines into completely autonomous and artificially intelligent machines. During the most recent disasters, swarms of aerial robots flew deep into the affected zones to become the primary source for damage surveying and aid delivery.

ML (machine learning) is a class of complex software applications that can learn and infer patterns from numbers, texts, images, videos, and other sources of information, and use the information for predicting outcomes for unseen scenarios. For example, an ML model that is designed for and properly trained using data from a river can predict and sound an alarm if the river reaches flooding conditions. ML also acts as the brain inside most robots and enables their decision making.

Land and Water Robots

In the aftermath of the 2011 earthquake and tsunami, Japan's leading experts from the Chiba Institute of Technology deployed their search-and-rescue robot "Quince" to assist emergency responders in searching for survivors under the rubble and survey the damages at the Fukushima Daiichi nuclear power plant (Frontiers of Engineering, 2014).

Quince can travel on almost any terrain: it has 4 independent legs with traction and its body also provides traction. At only 65 cm long, 48 cm wide and 22.5 cm tall, it can maneuver into tight spaces. It weighs 26.4 kg, so it can be transported to affected areas easily. Based on the desired application a selection of many attachments are available, such as wide-angle cameras, night vision cameras, microphones, speakers, two-way communication devices, as well as a gripper. Quince is also resistant to water, impact, and radiation (Fig. 3.1).

The U.S. Army Corps of Engineers used SeaBotix VLBV300, an underwater remotely operated robot manufactured in the US by Teletyne SeaBotix, to investigate bridge and seawall damage as part of the U.S. assistance to the Haitian government during the 2010 Haiti earthquake disaster (See Murphy, R., 2001, 2019).

Fig. 3.1 Quince land discovery robot

VLVB300 is SeaBotix's newest generation of remotely operated vehicles that can dive up to 300 m deep. It is available in many configurations for applications like port security, hull and infrastructure inspection, and light work such as object recovery and payload delivery (Fig. 3.2).

The above examples showcase scenarios where humans could not have reached easily, highlighting the biggest benefit from rescue robots; eliminating the risks related to injury and fatigue that rescuers take while entering affected zones along with providing access to hard to reach areas and scaling up rescue operation way beyond human capacity. For this reason, rescue robotics is gaining adoption in mining accidents, hostage situations, urban disasters, nuclear disasters as well as many other types of disasters.

Although rescue robotics development is focused on a wide range of applications like searching, reconnaissance, remapping, removing rubble, aid delivery, evacuation, and medical treatment, rescue robotics is still considered an emerging technology as there is a wide gap in application that still needs to be covered.

Fig. 3.2 SeaBotix underwater remotely operated robot

Aerial Surveillance and Aid Delivery Robots

Since 2014, Zipline, USA (a company pioneering delivery using drones) has been developing and deploying drones for aid delivery missions in Rwanda and Tanzania for providing aid and medications such as vaccines, HIV medication, blood, and IV tubes (Time, 2018).

Using drones, the relief organizations have been able to outpace ground-based supply transport as the drone delivery reaches a 50-mile radius in under an hour. Medical staff at clinics that have subscribed to drone deliveries can simply order new supplies by sending a WhatsApp text message. Drones' flight paths are determined using 3-D satellite maps in coordination with air traffic control and commercial airliners. The drone does not land at the delivery site, instead the package is released with a parachute and lands within a 5-meter wide landing zone. The cost of delivery via drone is reportedly comparable to conventional means if the areas are accessible via roads (Fig. 3.3).

Drones also have an important application in surveillance and damage assessment for accelerating aid recovery and rebuilding efforts. In the

Fig. 3.3 Zipline drone deploying a package

recovery efforts from the earthquake and tsunami that hit Indonesia on September 28th 2018, the Indonesian government was supported by the nonprofit disaster-relief agency "Team Rubicon Australia". The agency used aerial robotics to help Indonesian authorities assess damage and plan recovery operations (DroneDeploy Conference, 2019).

With the roads clogged with debris and bridges and airport runways destroyed, conventional means of access became very limited. Using ground crews for surveying would have taken a significant amount of time and exposed them to personnel risks. Aerial robotics offered a much speedier, safer solution; teams used drones to take photos and videos of given the destruction and inaccessible landscape.

"DroneDeploy" is a software company that offers a paid platform that manages all aspects of aerial mapping like, flight planning, flying, image processing, and map analysis. This platform allows relief organizations develop or expand their fleet using drones from any manufacturer while maintaining a common control and command interface. DroneDeploy's platform is particularly helpful when internet service is wiped out or

Fig. 3.4 Aerial map of affected areas

destroyed; it allows drones to fly entirely offline and later compress the retrieved imagery for processing using human-based or machine learning-based labeling. Labeling and analysis of imagery allow the teams to create aerial maps showing damage to buildings, roads, bridges, and other infrastructure. The drone data provide higher resolution imagery that supplements the satellite imagery, for a complete analysis and make data-driven decision making for a faster recovery process.

Figure 3.4 shows the aerial map generated in the DroneDeploy platform using data from a group of surveillance drones, without the need for experts to process the data.

AI as a Software Service During a Disaster

Unlike robotics which has existed for decades, machine learning is a relatively new branch of AI. Machine learning (ML) are algorithms that can perform a specific task without programmed instructions, instead the algorithm learns by deciphering patterns and inferences in the input data, hence they are also classified as AI. Machine learning has expanded pattern recognition far beyond human abilities. Accurate pattern recognition models have enabled better early warning systems (Journal reference: Water 2018, 10, 1536) and social media information-sourcing systems (23rd International Conference, Korea 2014).

AI has also enabled faster processing of insurance claims. After a disaster, insurance providers are inundated with claims which create delays while people are waiting on insurance payments to get their lives back in order. Insurance companies are using Natural Language Processing (AI) to help sift through claims data, to infer important information and to identify high-priority claims. This partially automates and streamlines the process for everyone involved, which enables faster processing of claims (Accenture article, 2018).

Early Warning Systems Let's examine the case of early warning systems for flood forecasting; an accurate flood forecasting system can help authorities move people out of harm's way and preemptively mobilize recovery resources. The forecast can also be used to trigger flood warning automatically, which will enable proper and on-time evacuation. However, predicting flooding with precise location and lead-time is complex, the current prediction systems are costly and slow. In addition, warnings are often ignored by authorities, as incorrect or un-actionable predictions have eroded trust in predictive systems. The ML-based prediction systems are proving to be a great success in flood forecasting, offering better accuracy, faster and more actionable insight, as well as being more cost efficient (Google AI – The Verge, 2019).

The cost of building and deploying a flood forecasting system can be quite high as to produce a reasonably accurate forecast, experts with in-depth knowledge of hydrological parameters analyze decades of data and build physics-based models that predict the future. By using ML algorithms, tasks such as dam monitoring can be automated, meteorological data analysis can be automated and administrative actions from system generated warnings can be automated, thus there are fewer overhead activities and costs. ML algorithms also simplify the prediction activity as they do not need as much temporal data or expert knowledge for interpretation, they develop their own inferences from the input data.

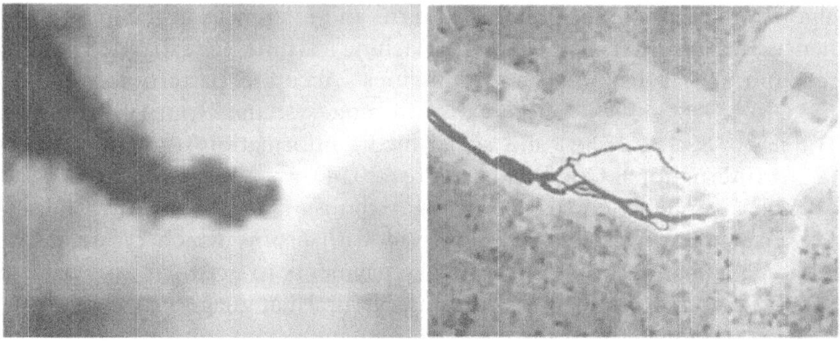

Fig. 3.5 Figure shows legacy flood model on the left and ML-based model on the right. (Google AI, 2018)

Figure 3.5, shows a flood simulation of the Musi river in Hyderabad, India. The image on the left side uses publicly available data while image on the right side uses Google data and Google's AI technology. The image on the right provides a much higher resolution of flood water flow, thus providing specific regions that can be evacuated rather than the entire city. Similarly, ML-based early warning systems can simulate other disasters at scale and map out regions with the highest probabilities of impact so that proper disaster management and mitigation strategies for the region can be developed, such as evacuation or sheltering policies and preparatory measures.

Information Sourcing Systems Social media platforms such as Facebook and Twitter are responsible for a surge in global connectivity. Low-cost and internet-enabled mobile devices have expanded connectivity to lesser developed parts of the world. These two trends have granted anyone the ability to broadcast detailed geo-tagged data and made crowdsourcing a predominant source for information gathering, however, it also poses the biggest risk for spreading misinformation.

In the recent September 2017 earthquake in Mexico, victims posted on social media platforms to ask for help as well as posting images of damage. AI was used to scrape information from millions of social media posts and provide clues to rescue workers about the hardest-hit areas and people in the most need, thus turning a historically one-way blind rescue operation into a two-way dialogue between connected crowds and smart response systems.

Artificial Intelligence for Disaster Response (AIDR) is pioneering this ML-based evaluation of data from social media platforms, they use natural language processing and computer vision techniques to analyze crisis information communicated on social media, content in different forms is consumed (e.g., text, image, video, etc.) to gain situational awareness and actionable information which is used by decision-makers, NGOs, affected communities, and scholars to improve the effectiveness of humanitarian strategies such as preparedness, mitigation, and response during natural disasters and other emergencies (Fig. 3.6).

Adoption of AI in Disaster Management

AI is a significant force multiplier in our ability to protect people and property in the face of disaster and is undoubtedly the future for disaster management, but to adopt AI in disaster management government organizations will need to complete a roadmap of prerequisites to ensure the adoption is efficient and effective.

1. At the center of AI is data, to make the best use of robotics and machine learning one needs the ability to gather and store data in an efficient manner, as well as develop enhanced techniques to collect various environmental parameters that characterize natural disasters. Data collection will need to include both normal and extreme conditions that lead to disasters. A high-level flow of data into predictions using machine learning is shown in Fig. 3.7.

2. Hiring and developing a team of AI experts focused on disaster management will provide targeted development in the field.
3. Government organizations should get early education and exposure to AI as a top-down adoption can provide proper guidance and a unified vision to all disaster management government agencies.

In Fig. 3.7, we see the flowing of past data into a statistical model that recognizes patterns and derives inferences. The patterns and inferences are stored as training results, and once a model is trained it does not require

Fig. 3.6 AIDR twitter feed of information during the earthquake in Pakistan (24 September 2019) (Explain which Pakistan earthquake?)

frequent retraining. A trained ML model uses training results to predict outcomes of previously unseen conditions (labeled as "unknown data").

Another major step in the adoption of AI is understanding its limitations. AI can make mistakes on instances, such as when the data it is trained

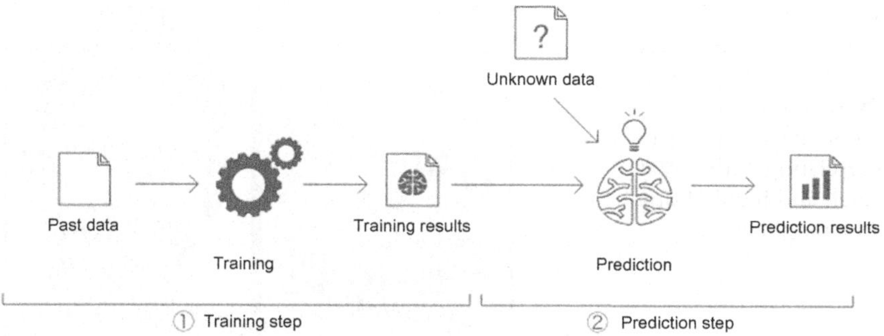

Fig. 3.7 Conversion of data into predictions using machine learning (www.magellanic-clouds.com)

on is flawed its predictions will be inaccurate. However, if we get too reliant on AI and forgo due diligence we might ignore the step of cross-validation, instead directly acting on incorrect predictions which can have significant consequences. Therefore, a team of AI experts needs to constantly evolve and validate the technology to ensure their reliability and fit for real-life scenarios.

Another limitation of AI is that it makes predictions based on past records of natural disasters, the current state-of-the-art machine learning models are unable to consider things like climate change trends on the future impact of natural disasters, thus current models are only suitable to making shorter-term predictions.

Lastly, most technology, prediction and output from robotics and AI are algorithmic which can't be designed to be totally unbiased. Rescue robots and ML algorithm might develop a data-based bias, which can then lead to different actions for different individuals in similar situations. Julia Angwin, Larson, Mattu, and Kirchner (2016) in their study highlighted an example of AI generated bias, Fig. 3.8 shows racial bias in threat detection AI software, you can see in the results that incorrect threat labels have been applied considering the individual's criminal history. In this case the predication algorithm has inferred an inherent threat with the race and is biasing its labels for individuals regardless of their criminal history, this makes human judgement critical, as in such cases a human would have been able to use judgment and unbias the actions or output from AI.

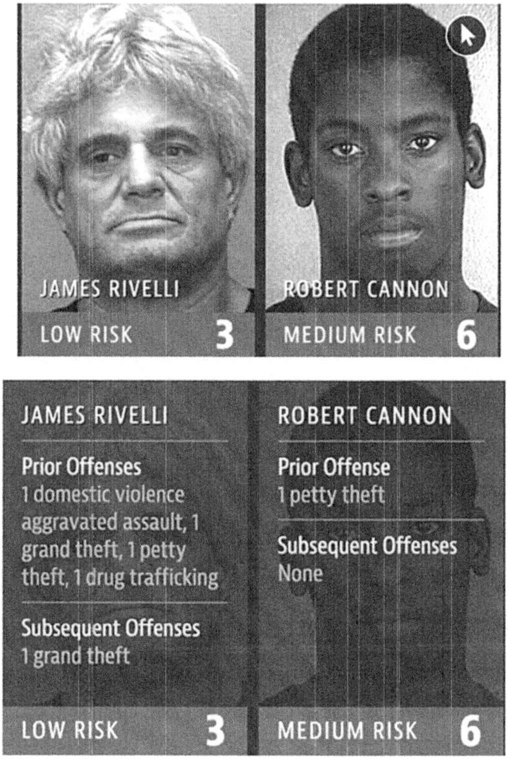

Fig. 3.8 COMPAS Software results by Julia Angwin et al. (2016), show results where criminal history was not correlating with threat label, and highlighting AI's racial bias

So it is critical the robotics and ML be governed by a human command, and human judgement still remains an integral part even as AI gets adopted in disaster management.

Conclusion

This chapter explained that AI opens up new possibilities to scale disaster management in ways that did not exist before. Although AI in disaster management is still largely a research field, we reviewed many companies pioneering new products and services that are ready for adoption. With the use of robotics, recovery and survey operation are being expanded while the risk of injury to rescue crew is being reduced, and aid is being delivered swiftly deep into remote locations.

Machine learning has enabled us to understand large amounts of data without having expert knowledge of the data itself, it is being used to develop more efficient and accurate early warning systems by improving their prediction accuracy and providing more advance notice before an event. ML has provided the ability to simulate disasters at scale to create maps of regions that are expected to impact.

AI is improving the recovery process, mapping damage to infrastructure, and proving direction to rescue efforts. It is also helping with expediting the claim process by providing automation of tasks for insurance companies.

Although AI will modernize disaster management, its implementation needs to be directed and governed by a body to experts to detect and prevent the creep of data-based biases.

References

Angwin, J., Larson, J., Mattu, S., & Kirchner, L. (2016). *Machine Bias*. Available at https://www.propublica.org/article/machine-bias-risk-assessments-in-criminal-sentencing. Accessed 17 Aug 2017.

Baker, Aryn. The American Drones Saving Lives in Rwanda. *Time*. Retrieved October 12, 2018.

DroneDeploy Conference. (2019).

DroneDeploy Blog. https://blog.dronedeploy.com/drones-assess-the-aftermath-of-a-indonesias-destructive-earthquake-1e60611d0abd

Frontiers of Engineering. (2014). Tomoaki Yoshida: Quince: Monitoring Robot for Disaster Response in the Fukushima-Daiichi Power Facility.

Imran, M., Castillo, C., Lucas, J., Meier, P., & Vieweg, S. (2014). AIDR: Artificial Intelligence for Disaster Response. https://doi.org/10.1145/2567948.2577034

Machine Learning in Insurance. (2018). https://www.accenture.com/_acnmedia/pdf-84/accenture-machine-leaning-insurance.pdf

Mosavi, A., Ozturk, P., & Chau, K.-W. (2018). Flood Prediction Using Machine Learning Models: Literature Review. *Water, 10*, 1536.

Murphy, R. R. (2001). *Disaster Robotics*. Cambridge, MA: MIT Press.

Murphy, R. R. (2019). *Introduction to AI Robotics*. Cambridge, MA: MIT Press.

Nevo, S., Anisimov, V., Elidan, G., El-Yaniv, R., Giencke, P., Gigi, Y., Hassidim, A., Moshe, Z., Schlesinger, M., Shalev, G., Tirumali, A., Wiesel, A., Zlydenko, O., & Matias, Yossi. (2019). *ML for Flood Forecasting at Scale*. https://www.blog.google/products/search/helping-keep-people-safe-ai-enabled-flood-forecasting/

Noymanee, J., Nikitin, N., & Kalyuzhnaya, A. (2017). Urban Pluvial Flood Forecasting Using Open Data with Machine Learning Techniques in Pattani Basin. *Procedia Computer Science, 119*, 288–297. https://doi.org/10.1016/j.procs.2017.11.187

Sud, K. (2018). Role of Heavy Machinery in Disaster Management. In A. Singh, M. Punia, N. Haran, & T. Singh (Eds.), *Development and Disaster Management*. Singapore: Palgrave Macmillan.

CHAPTER 4

Optimal Visual Cues for Smartphone Earthquake Alert Systems: Preliminary Data from Lab and Field Experiments

Eran Lederman, Tomer Shemi, Noga Ensenberg-Diamant, Lior Shalev, Amir Hazan, Naama Marcovits, Yona Weitz, Or Haklai, Tal Badichi, Bar Segal, Dan Fishbein, and Hillel Aviezer

It's 9 pm and you are enjoying a glass of good wine and the company of friends in your living room. Suddenly, your cellphone sends an alert and you realize you have several seconds before a major earthquake hits your town. Without delay, you rapidly run for cover outside. Although your home is severely damaged—you and your loved ones are saved. Several

E. Lederman (✉) • T. Shemi • Y. Weitz • O. Haklai • T. Badichi • B. Segal • D. Fishbein
Bezalel Academy of Arts and Design, Jerusalem, Israel

N. Ensenberg-Diamant • H. Aviezer
Hebrew University of Jerusalem, Jerusalem, Israel

L. Shalev • A. Hazan • N. Marcovits
eVigilo, Englewood Cliffs, NJ, USA

T. V. V. Kumar, K. Sud (eds.), *AI and Robotics in Disaster Studies*, Disaster Research and Management Series on the Global South,
https://doi.org/10.1007/978-981-15-4291-6_4

years ago, such a scenario may have seemed like pure science fiction but recent advances in early detection of seismic P waves (before the destructive S waves) have rendered it a reality (Allen & Kanamori, 2003). For example, at an August 2014 earthquake in California's Napa county, people received an alert five seconds before the destructive magnitude-6.0 shaking begun (Gerber, 2014).

While the geo-technological advances in earthquake detection are capable of sending an alert to end-user's mobile phones, the physical nature of the alert has received little attention by government agencies and has undergone little development over past years. For example, the alerting system "ShakeAlert" (Burkett, 2014) developed by the U.S. Geological Survey (USGS) includes a combination of auditory warnings and graphical displays while the interface of the US national alerting system includes a text message with an auditory cue—in both cases, it is unclear if these specific characteristics are indeed optimal for the human end-user (see Fig. 4.1). While the specific characteristics of the alert may seem inconsequential for natural disasters that take hours to build up they are utterly critical in the case of earthquakes because of the extremely short duration between the alert onset and the event.

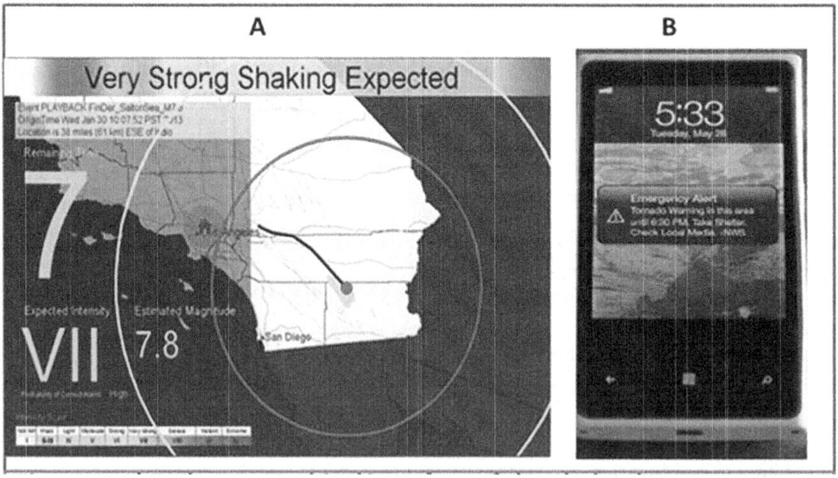

Fig. 4.1 Examples of end-user interfaces in alert systems apps. (a) "ShakeAlert" system. (b) National emergency alerting system (EAS)

Aside from comparing between classic alert visual stimuli such as text (e.g., the word "DANGER") and hazard icons (e.g., ⚠), we were further interested in a potentially new class of alert stimuli, namely, that of social-emotional nature. Emotional signals such as facial expressions are a primitive and evolutionary based signaling system of affective relevance (Bänziger & Scherer, 2010; Schmidt & Cohn, 2001). The emotional value of facial expressions is recognized universally and cross culturally and they are often considered a biologically hardwired and evolutionary-adaptive signal (Ekman, 1972, 2007; Elfenbein & Ambady, 2002; Marsh, Elfenbein, & Ambady, 2003; Sauter, Eisner, Ekman, & Scott, 2010; Susskind et al., 2008). Because of the natural inherent importance of facial expressions, they serve as optimal stimuli for conditioning paradigms. For example, electric shocks are rapidly and readily conditioned to fearful faces but less so to happy faces (Ohman, 2000).

The inherent biological significance of facial expressions can be demonstrated in humans from a very early age. Four-month old infants can discriminate facial expressions of anger, fear and surprise (Serrano, Iglesias, & Loeches, 1992). At the age of twelve months, infants successfully use information about an adult's direction of gaze and emotional expression to predict adult action and they efficiently utilize negative expressions of caregivers as danger cues (Mumme, Fernald, & Herrera, 1996).

Facial signals, in particular negative expressions, are highly potent attractors of attentional processing even when presented outside the scope of awareness. In fact, negative facial expressions are recognized and oriented to even among visual neurological patients in whom spatial awareness is impaired due to stroke (Vuilleumier & Schwartz, 2001). Experimental methods which allow presenting stimuli without visual awareness such as continuous flash suppression (Yang, Zald, & Blake, 2007) and binocular rivalry (Yoon, Hong, Joormann, & Kang, 2009) also demonstrate a superiority for facial expressions in emerging out of unconscious processing and breaking into awareness.

In addition to our interest in the potential utility of facial expressions as "attention grabbers", faces may hold additional advantages for alerting systems. Unlike traditional alert systems that simply bombard the end-user with info and "throw him in the water" on his own. A social alerting system can actually serve as an agent that guides the end-user through the process, potentially serving as a source of motivation, encouragement and guidance. This may give faces an advantage as users may socially connect with the alerting interface, improving their experience of trust and perceived safety when using the app.

Contemporary studies and innovative technologies are already able to identify individual people within a large audience by means of face photographs only. It is clear that face recognition and data mining technologies are increasingly being improved by the use of big data and artificial intelligence.

Examples of the use of these technologies can be seen in the software developed by FACE2GEN that uses facial phenotypes to facilitate comprehensive and precise genetic evaluations or by FindFace software capable of detecting faces, emotions, gender and age of people by analyzing their photos only.

While the above examples demonstrate how technology can aid at the data collection from faces, faces may also be efficient as an alerting stimulus. Our side in this study deals with how to improve and correct the cellular alert for an earthquake, first and foremost in the attention capture, and the ability to motivate people to act quickly and efficiently in light of the danger they face.

Experiment 1. Optimal Stimuli for Capturing Peripheral Attention

One challenge to alert systems is that they must attract user's attentions while placed outside the center of visual attention. Consider the case of Jane who is deeply immersed in her favorite Netflix show playing on TV while her smartphone is placed on the sofa beside her. Jane's reaction to information appearing on her phone (e.g., an email notification) may be delayed simply because her focus of visual attention is elsewhere.

In experiment 1 we created an ecologically valid experimental setting to test which stimuli would best capture participant's attention while they are engaged in watching a comic video. Specifically, participants were requested to detect numbers presented together with alerting stimuli (text, icons and fearful faces) presented on a separate screen, peripherally to the main task video. Our dependent variable was the reaction time to responding to the number.

Methods

Participants 30 students (18 females, Mean age = 24) from the Hebrew University participated in the experiment for course credit or payment.

Stimuli and Materials The main task for participants involved watching an 18 minute "best bits" YouTube video of the sitcom *Friends*. Alongside the main task, a concurrent "attention capture task" took place. Stimuli for the attention capture task included alert faces, alert icons and alert texts. Alert faces included 10 facial expressions of fear from the ADFES database (Hawk, Van der Schalk, & Fischer, 2008). Alert icons included 10 graphic icons conveying danger (e.g., skull and cross bones). Alert texts included single word exclamations such as "Danger".

All stimuli were equated for size and average visual angle.

Procedure The main task video clip was presented at full screen size on an 18 inch ThinkPad laptop placed directly in front of the participants at a distance of ~60 cm. They were instructed to observe the clips carefully as they would be asked to answer questions regarding the storyline at the end of the experiment. While watching the video, a concurrent attention capture task took place to the right or left side of participants. A second computer screen was placed at a distance of 50 cm, turned slightly towards the participant. Three categories of alerting stimuli (faces, icons and text) were randomly presented on this peripheral screen every 30–90 seconds. In order to keep participants as naïve as possible, they were told that a number with a stimulus above it will be occasionally presented. Their task was to detect the number and press the key corresponding to the presented number as quickly as possible. While the video ran uninterrupted during the entire task, the numbers and alert icons appeared on screen for 1000 ms on each trial, and were then removed. No feedback was given for the attention side task. Upon completion of the experiment, participants were given a brief quiz about the video. This was intended to confirm that attention was adequately located to the main task.

Results All participants successfully answered the questions about the video, indicating that their attention was adequately attended to the content of the video. Our main dependent variable was the RT to detection of the numbers on the peripheral screen. To this end, the mean reaction times to correct pressing of the numbers keys was compared across the 3 conditions. Figure 4.2 presents the RT for the pressing of correct numbers in each of the alert conditions. A 3-way repeated measures ANOVA run on the mean RT data indicated a significant effect of the condition [$F(2,58) = 6.6$, $p < 0.003$, $\eta_p^2 = 0.18$]. Paired t-tests across conditions indicated that the RT in the text condition was slower than in the icon condition ($t(29)=3.7$, $p<0.001$ and slower than in the face condition ($t(29)=2.4$, $p<0.023$). However, no significant difference was found between the graphic alert icons and between the faces ($t(29)=0.9$, $p=0.34$).

Discussion Experiment 1

The results of experiment 1 demonstrated more efficient attention capture when alert icons and faces were presented than when alert text was. However, icons and faces were equally deficient in capturing attention.

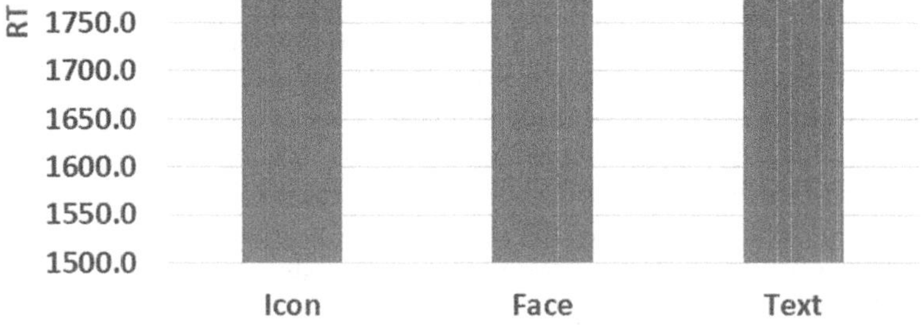

Fig. 4.2 Reaction time (RT) results of experiment 1. Icons and faces resulted in faster responses than text

This finding may be important given the popular prevalence of text in many alert apps, which may be suboptimal. In experiment 2 we continued to examined compare alert faces and icons in a more real-life setting. Unlike lab experiments in which multiple stimuli are repeated many times, real-life alerts are a one-shot, unexpected event. It was thus imperative to examine the alert value and accompanied psychological perceptions of faces and icons in a more ecological field-study setting.

Experiment 2. Optimal Stimuli, Personalization and Practice in a Real-Life Field Experiment

In order to examine optimal alert interfaces in real-life ecological settings, we conducted a controlled field experiment. This allowed us to examine the efficiency of reactions to different alert stimuli and the optimal personal experience and learning process associated with such stimuli in a real-life setting. Specifically, the study examined the impact of several factors: stimuli of alert (icon, face), the process of personalization (choosing one's icon/face vs. having it assigned randomly), and practice (with or without).

Based on study 1, we did not expect a main effect of the stimuli as both proved equally efficient in capturing attention. We did, however, predict a main effect of practice as this effect was established in multiple studies (e.g., Johnston et al., 2011). With regard to personalization, we predicted an interaction: due to the special status of faces in social, interpersonal interactions we expected that personally choosing one's alert would enhance performance more strongly for faces than for icons. Humans are highly acquainted with the social task of analyzing the attractiveness, expressiveness and traits read from faces. As such, choosing a favorite face for an alert stimulus may create a more meaningful and engaging experience than choosing a graphic icon.

In addition to examining the motor aspect of the alert response (i.e., RT to reactions), we were also interested in the psychological reactions participants developed towards different stimuli and conditions. Specifically, we predicted that participants would experience more positive experiences (e.g., feeling more confident, safe and trusting of the alert) when alerted by a face than by an icon. This would be the case due the personal feelings and social attachment that may arise when encountering faces vs. icons, the latter being non-personal by nature.

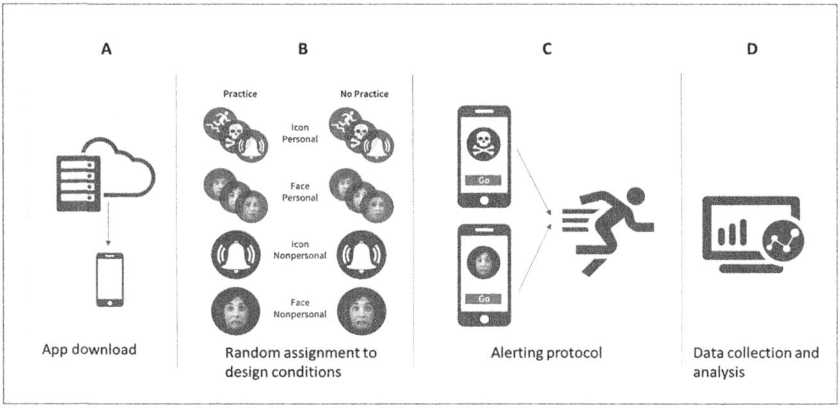

Fig. 4.3 Illustrative layout of the experimental design and procedure of experiment 2

Methods

Participants 220 participants were recruited from an Israeli online participant pool (Panel4all) and compensated with ~10$US for completing the study.

Stimuli 3 faces and 3 icons were chosen from those used in experiment 1. These visual stimuli appeared together with an auditory alert sound and a vibrating sensation. Thus all conditions were conducted with identical auditory and vibration alerts which were intended to enhance the likelihood of response.

Design The design was a 2 (stimuli: face, icon) x 2 (personalization: yes, no) x 2 (practice: yes no), between subject design. See Fig. 4.3 for the experimental design and procedure layout (Fig. 4.3).

Task and Procedure

Participants were randomly assigned to the different conditions and downloaded the app via a link received through the online subject pool. After reading the basic instructions and giving consent, participants

Fig. 4.4 Example of the app layout and alert procedure. Left image portrays the stage of personalization of the face alert image. The right image portrays the surprise alert

were exposed to 3 stimuli on the screen (3 icons or 3 faces). In the personalization condition, subjects were given 10 seconds during which they were requested to choose one of the 3 stimuli which they would like to appear as their alert stimuli. The choice was indicated by pressing on the screen. In the non-personalization condition, the 3 stimuli (3 icons or 3 faces) appeared for 10 seconds and one of the 3 stimuli was randomly chosen by the app. See Fig. 4.4 for an illustration of the app layout stages and alert procedure.

Next, depending on their random allocation, participants in the practice condition received a practice trial in which an alert stimulus appeared on the screen with a number below it. At the bottom of the screen was a panel with several numbers and subjects were requested to match the number below the stimuli, with the corresponding number in the panel. Immediately after the number matching, subjects were requested to walk 10 steps and press a "SAFE" button indicating that they completed the task. This practice was essentially identical to the actual alert that subjects would be exposed to during the actual experiment. Participants in the no-practice group did not go through with this procedure.

In order to invoke and disseminate the experiment, we collaborated with eVigilo, a tech company specializing in mass-notification of cellular alerts. The technology that was delivered by eVigilo for this experiment, was based on eVigilo product Smart-eVigilo, an Emergency Mass-Notification and Alert multi-channel solutions, where its Server side send the messages and collect the response, and its application based product served as the end user side of the experiment. Smart-eVigilo is a Web based mission critical system, designed to operate under extreme conditions of emergency in order to save lives.

Using eVigilo's technology, a week after enrolling, participants were mass alerted by surprise with the initiation of the actual alerting procedure. The alert was triggered around 18:00 pm and mass disseminated to all subjects simultaneously. Subjects had no prior knowledge regarding the date or time when the alert would occur. As noted, the actual alert was identical to the practice procedure and responses were recorded by the app.

Following the completion of the app alert task, participants were diverted to the online panel website and were requested to complete a brief survey about their user experience during the alert. In the survey, participants were requested to indicate with a 1–7 scale how safe they felt during the alert, and to what degree they would trust the app in case of a real emergency. Subsequently, they were thanked for their participation and compensated.

Results

Data Preparation Due to the uncontrolled, field nature of the experiment, a relatively large proportion of subjects did not respond in a timely manner to the alert (e.g., subjects that were driving at the time of the alert, subjects who did not have their phones next to them, etc.). A data quality report from eVigilo (who collected the data their servers) indicated subjects with invalid or corrupt data logs, subjects that had not pressed the "SAFE" button, and subjects that failed to respond to the alert altogether. Among subjects that did complete the experiment, a cutoff of 120 sec was used to remove people that responded very slowly—likely due to not seeing or being able to respond in a timely manner. After exclusion, we remained with an N =155. The N within the different groups ranged from 49 to 12. Given the variance and the low N in some of the groups, the statistical power of our experiment was low. We therefore opted for a lenient threshold and pursued marginal interaction trends to increase the likelihood of detecting effects in the data. Consequently, the findings should only be viewed as exploratory and should be interpreted with caution, awaiting an additional large-scale replication.

Results

Reaction Time A 2 (stimuli: face, icon) x 2 (personalization: yes, no) x 2 (practice: yes no), between subject ANOVA was run on the RT data (data reported in seconds) A significant main effect of practice indicated that participants, were overall faster with practice (M=21.8, SD=14.4) than without practice (M=31.8, SD=23), $F(1,147)=8.93$, $p < 0.001$. Additionally, a significant 3 way stimuli x personalization x practice interaction was revealed, $F(1,147)=5.5$, $p < 0.02$.

In order to characterize the 3 way interaction, we ran separate stimuli x personalization ANOVA's for the practice and no practice conditions. Within the practice condition, a marginal stimuli x personalization interaction was found, $F(1, 57)=3.3$, $p=0.07$. A trend was found indicating that faster RT's were associated with personalization (i.e., choice) of faces as well as with non-personalization of icons. Within the no-practice condition, a marginal stimuli x personalization interaction was found, $F(1, 90)=3.4$, $p=0.06$. A trend was found indicating that faster RT's were associated with personalization of icons, as well as with the non-personalization of faces.

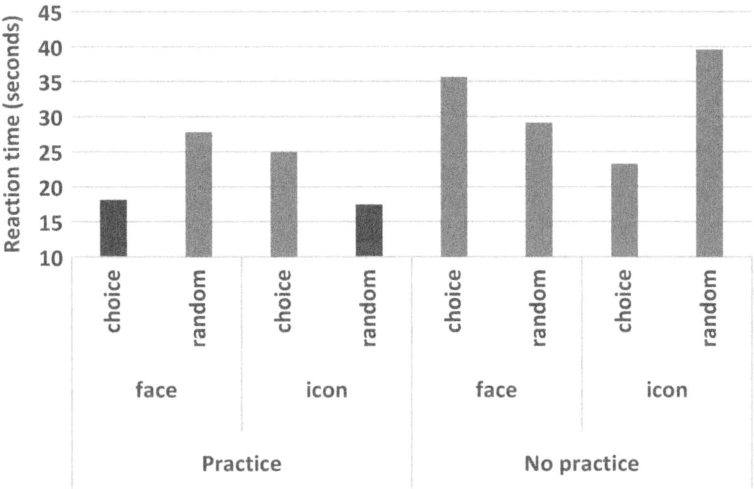

User Experience Ratings

In Case of a Real Earthquake, How Likely Would You Be to Trust This App?

A practice x stimuli interaction emerged F(1,147)=6.3, p<0.01, indicating that with practice, faces received a higher degree of trust than icons. By contrast, without practice, an opposite pattern emerged.

In Case of a Real Earthquake, How Safe Would You Feel with This App?

A practice x stimuli interaction emerged F(1,147)=6.5, p<0.01, indicating that with practice, faces received a higher degree of safety ratings than icons. By contrast, without practice, an opposite pattern emerged.

To What Degree Did I Feel Safe During Walking During the Alert?

A trend for a practice x personalization interaction emerged F(1,114)=3.1, p=0.78, indicating that with practice, personalization resulted in a higher degree of safety ratings than non-personalization. By contrast, without practice, an opposite pattern emerged.

Overall Mood During the Experiment

No main effects or interactions approached significance for the ratings of overall mood (significance levels ranging from $p=0.24$ to $p=0.78$), suggesting that the previously reported findings in user experience did not results from generalized mood effects, but rather from specific differences in user experience.

Discussion Experiment 2

Experiment 2 offers a unique field approach to studying optimal alert stimuli and procedures. While the study is notably underpowered, some interesting conclusions may be tentatively deduced from the data, awaiting future replication and extension. First, the robust beneficial role of practice in alert compliance and performance emerged in our findings. Any alert application being considered in the future should likely consider this important factor. Second, our results suggest that personalization, i.e., one's ability to choose their own alert, is only beneficial for face stimuli, but not for classic graphic icons of alert. While the personalization process takes additional time, it may well be worth the effort. Participants consistently rated the face alerts as more trustworthy and enhancing of a safe user experience. Importantly, faces may induce a more interpersonal relation between users and the app, thereby increasing compliance and improving performance.

General Discussion

Earthquake alerts allow participants several seconds (~10) to rapidly realize the alert is taking place, and to act (e.g., get out of a building). With such brief time intervals, literally every second counts. To date, alert systems have focused on the mere notification of subjects, with little thought devoted to the medium and procedure of this alert. However, our lab and field studies strongly suggest that developers can optimize their alert systems by choosing the right kind of visuals, allowing the appropriate personalization with these visuals, and having subjects practice the response protocol.

In experiment 1 we demonstrated the superior attention capture when alert icons and faces were presented than when alert text was. The use of text is surprisingly common in many alert systems. Aside from the limitation of literacy, text may be suboptimal for capturing attention. Having established the advantage of visual cues in capturing attention, Experiment 2 continued to compare alert faces and icons in a more real-life setting. Overall our preliminary results hint to possible future protocol guidelines. Aside from the obvious importance of practice our results suggest that personalization is only beneficial for face stimuli. Most importantly, participants consistently rated the face alerts as more trustworthy and enhancing of a safe user experience. Considering the high drop out and attrition rates that many apps have, establishing a personal social bond with a face avatar may increase usability and performance.

These novel results shed important light on the optimal underlying factors of efficient alert systems and may help design such systems in the future. Artificial intelligence still needs to bridge many gaps with regard to creating effective tools for generating high-quality personal cellular alerts. In particular, the ability to use artificial intelligence in order to collect data and design optimal alerts customized for the unique needs of the user may be a promising and exciting direction. User habits, preferences, emotional state and personality may all be taken into account after obtained from user data. While such a vision seems of great importance, the research and technology have not yet reached the level of maturity that will enable the creation of an accurate personalized cellular alert. However, such gaps are minimizing rapidly and technologies available today seemed unimaginable several years ago.

As noted, we view our studies as a first tentative step towards more established research in the field. Aside from significant increases in power that our study (especially experiment 2) suffers from, many important questions remain unanswered. In both experiments, the auditory alert was held constant either by omitting it (experiment 1) or by using a constant alert sound across all conditions (experiment 2). The acoustic alert is without doubt a critical component of any alert, but little is known about how it interacts with the visual cue. Given the advantage of face stimuli found in our work, future studies may compare the utility of human alert voices vs. artificial alert sounds (beeps/sirens).

An additional direction not covered here is that the potential of the personalization process is probably much more complex than offered in our limited design. Different personality individuals may benefit from choosing a face avatar that best suits their liking. Factors such as race, gender, attractiveness and personality attributes of the alert face may all contribute to the ideal alert system. Looking at the more distant future, user characteristics obtained with consent may be analyzed and used to improve the alert procedure and medium. For example, practice is important, but over-practice may lead to burnout. A system that customizes and optimizes the practice routine to the unique characteristics of the user may improve the alerts efficiency.

In our vision, the role of the alert app extends far beyond the initial activation of participants. Rather, the app should continue to serve as a companion to the user throughout the entire event. This may include the process of taking cover, being trapped, seeking for loved ones and friends, and assessing safe harbors. In all these processes, expressing dynamic emotions from the app, and reading expressions from the user may be critical. While futuristic, we believe these directions reflect current trends. However, only additional research will establish which protocols are ideal for optimizing the alerting process.

Acknowledgments This study was supported by a grant (#47953) from the Israeli Ministry of Science, Technology & Space.

References

Allen, R. M., & Kanamori, H. (2003). The Potential for Earthquake Early Warning in Southern California. *Science, 300*(5620), 786–789.

Bänziger, T., & Scherer, K. R. (2010). Introducing the Geneva Multimodal Emotion Portrayal (Gemep) Corpus. *Blueprint for Affective Computing: A Sourcebook*, 271–294.

Burkett, A. (2014). ShakeAlert—An Earthquake Early Warning System for the United States West Coast. http://pubs.usgs.gov/fs/2014/3083/pdf/fs2014-3083.pdf

Ekman, P. (1972). Universals and Cultural Differences in Facial Expressions of Emotion. In J. Cole (Ed.), *Nebraska Symposium on Motivation* (Vol. 19, pp. 207–283). Lincoln, NE: University of Nebraska.

Ekman, P. (2007). *Emotions Revealed: Recognizing Faces and Feelings to Improve Communication and Emotional Life*. Macmillan.

Elfenbein, H. A., & Ambady, N. (2002). On the Universality and Cultural Specificity of Emotion Recognition: A Meta-Analysis. *Psychological Bulletin, 128*(2), 203.

Gerber, M. (2014). After 6.0 Earthquake, Flames Light Up Napa Mobile Home Park. *Los Angeles Times*.

Hawk, S. T., Van der Schalk, J., & Fischer, A. H. (2008). *Moving Faces, Looking Places: The Amsterdam Dynamic Facial Expressions Set (ADFES)*. Paper presented at the 12th European Conference on Facial Expressions, Geneva, Switzerland.

Johnston, D., Tarrant, R., Tipler, K., Coomer, M., Pedersen, S., & Garside, R. (2011). Preparing Schools for Future Earthquakes in New Zealand: Lessons from an Evaluation of a Wellington School Exercise. *The Australian Journal of Emergency Management, 26*(1), 24.

Marsh, A. A., Elfenbein, H. A., & Ambady, N. (2003). Nonverbal "Accents" Cultural Differences in Facial Expressions of Emotion. *Psychological Science, 14*(4), 373–376.

Mumme, D. L., Fernald, A., & Herrera, C. (1996). Infants' Responses to Facial and Vocal Emotional Signals in a Social Referencing Paradigm. *Child Development, 67*(6), 3219–3237.

Ohman, A. (2000). Fear and Anxiety: Evolutionary, Cognitive, and Clinical Perspectives. In M. L. J. M. Haviland-Jones (Ed.), *Handbook of Emotions* (2nd ed., pp. 573–691). New York: Guilford Press.

Sauter, D. A., Eisner, F., Ekman, P., & Scott, S. K. (2010). Cross-Cultural Recognition of Basic Emotions through Nonverbal Emotional Vocalizations. *Proceedings of the National Academy of Sciences, 107*(6), 2408–2412.

Schmidt, K. L., & Cohn, J. F. (2001). Human Facial Expressions as Adaptations: Evolutionary Questions in Facial Expression Research. *American Journal of Physical Anthropology, 116*(S33), 3–24.

Serrano, J. M., Iglesias, J., & Loeches, A. (1992). Visual Discrimination and Recognition of Facial Expressions of Anger, Fear, and Surprise in 4-to 6-Month-Old Infants. *Developmental Psychobiology, 25*(6), 411–425.

Susskind, J. M., Lee, D. H., Cusi, A., Feiman, R., Grabski, W., & Anderson, A. K. (2008). Expressing Fear Enhances Sensory Acquisition. *Nature Neuroscience, 11*(7), 843–850.

Vuilleumier, P., & Schwartz, S. (2001). Emotional Facial Expressions Capture Attention. *Neurology, 56*(2), 153–158.

Yang, E., Zald, D. H., & Blake, R. (2007). Fearful Expressions Gain Preferential Access to Awareness during Continuous Flash Suppression. *Emotion, 7*(4), 882–886.

Yoon, K. L., Hong, S. W., Joormann, J., & Kang, P. (2009). Perception of Facial Expressions of Emotion during Binocular Rivalry. *Emotion, 9*(2), 172.

CHAPTER 5

Using Artificial Intelligence and Social Media for Disaster Response and Management: An Overview

Ferda Ofli, Muhammad Imran, and Firoj Alam

Introduction

During disasters and emergencies, humanitarian organizations, government agencies, public health authorities, and military are tasked with responsibilities to help affected and vulnerable populations (Gralla, Goentzel, & Van de Walle, 2013). These formal response organizations rely on timely and credible information to make rapid decisions to launch relief operations. The information needs of these stakeholders vary depending on their role, responsibilities, and the circumstances that they deal with (Vieweg, Castillo, & Imran, 2014). However, during time-critical situations, the importance of timely and factual information increases, especially when no information from other traditional sources such as TV or radio is available (Castillo, 2016; Vieweg, 2012).

F. Ofli (✉) • M. Imran • F. Alam
Qatar Computing Research Institute, Hamad Bin Khalifa University, Doha, Qatar
e-mail: fofli@hbku.edu.qa; mimran@hbku.edu.qa; fialam@hbku.edu.qa

T. V. V. Kumar, K. Sud (eds.), *AI and Robotics in Disaster Studies*, Disaster Research and Management Series on the Global South,
https://doi.org/10.1007/978-981-15-4291-6_5

The growing use of mobile technologies and social media platforms such as Twitter and Facebook has provided general public with practical and effective opportunities to disseminate and ingest information. Millions of people increasingly use social media to share information during natural or human-induced disasters (Castillo, 2016; A. L. Hughes & Palen, 2009; Purohit et al., 2014). For instance, victims of several recent hurricanes, including Hurricane Harvey, Hurricane Irma, and Hurricane Florence, have turned to social media to request help and assistance, as emergency hotlines were sometimes unreachable due to the high volume of calls (Frej, 2018; MacMillan, 2017; Rhodan, 2017). According to Rhodan (2017), a woman with two children were rescued during Hurricane Harvey after the woman posted a message asking for help on Twitter: "I have 2 children with me and tge [*sic*] water is swallowing us up. Please send help." Similarly, an image showing elderly residents sitting in greenish flood water was shared on Twitter more than 4800 times, which elicited an urgent help being dispatched to the nursing home (Rhodan, 2017).

Research studies have demonstrated the usefulness of social media information for a variety of humanitarian tasks to increase "situational awareness" (e.g., gaining insights into the situation as it unfolds, identifying reports about missing and found people, understanding urgent needs of the affected communities, and assessing the severity of damage) (Starbird, Palen, Hughes, & Vieweg, 2010; Vieweg, 2012). Although, information available on social media could be useful for response agencies, making sense of it under time-critical situations is a challenging task (Hiltz & Plotnick, 2013). For instance, due to high-volume and high-velocity of social media data streams, manual analysis of thousands of social media messages is practically impossible (Hiltz, Kushma, & Plotnick, 2014; Ludwig, Reuter, Siebigteroth, & Pipek, 2015).

Making sense of social media data to help responders involves solving multi-faceted challenges including parsing unstructured and brief content, filtering out irrelevant and noisy content, handling information overload, among others. Over the last few years, a number of artificial intelligence (AI) techniques and computational methods have been proposed to extract useful information from social media data for disaster response and management under time-critical situations (Stieglitz, Bunker, Mirbabaie, & Ehnis, 2018a; Stieglitz, Mirbabaie, & Milde, 2018c). These techniques aim to solve various challenges ranging from information filtering, overload, and categorization to summarization (Castillo, 2016; Imran, Castillo, Diaz, & Vieweg, 2015; Rudra et al., 2016).

The most important factor that drives scientific progress is the availability of public datasets, preferably with ground-truth annotations, collected from social media platforms. The most prominent datasets in the crisis informatics domain include CrisisLex (Olteanu, Castillo, Diaz, & Vieweg, 2014) and CrisisNLP (Imran, Mitra, & Castillo, 2016). In addition to having been widely used in the literature, these resources have quickly become online repositories for other crisis-related datasets. Although the focus has been almost exclusively on textual data analysis, Alam, Ofli, and Imran (2018a) have recently addressed this limitation by publicly releasing a multimodal Twitter dataset with human annotations for several humanitarian tasks. It is important to note that these publicly available datasets do not only facilitate development of new algorithms and solutions but also allow for benchmarking new algorithms and solutions against the state of the art, and hence, measure the scientific advancement.

While social media can help increase situational awareness, inform rescue operations, and save lives, its value is highly unexploited by response teams, in part due to the lack of tools that can help filter actionable information from the big data posted on social media during disasters. A recent study by Villegas, Martinez, and Krause (2018) reported that during Hurricane Harvey, the US Federal Emergency Management Agency (FEMA) missed 46% of the critical damage information posted by affected individuals on Twitter, and thus, many areas heavily impacted by the hurricane were missed from the original damage estimates provided by FEMA despite the fact that several studies have already shown that rapid early damage assessment can be achieved by analyzing social media data (Enenkel et al., 2018; Kryvasheyeu et al., 2016; Nguyen, Ofli, Imran, & Mitra, 2017c). This observation underlines the need for developing tools that can identify critical information from social media data in real-time and incorporating such tools into the operational decision-making routines of the response organizations. To that effect, a number of publicly available tools have been developed to collect and automatically process social media data during disasters (Burel, Saif, Fernandez, & Alani, 2017; Imran, Castillo, Lucas, Meier, & Vieweg, 2014; Meier, 2012; Okolloh, 2009). The common features of these systems include classifying tweets, grouping them into clusters, visualizing them into timelines, geotagging them onto maps, and visualizing topics, sentiment, and concepts over time.

In summary, to tackle this big data challenge in the humanitarian context, the scientific community, in close collaboration with governmental and non-governmental organizations, has undertaken the responsibility to curate large-scale datasets, to develop new computational models, and to build necessary systems and services. There have been significant progresses in all three dimensions in less than a decade. In this paper, we provide an overview of the state-of-the-art datasets collected, methodologies developed, and systems built in the disaster response and management domain based on artificial intelligence and social media.

Research

In this section, we first provide an overview of social media-driven crisis informatics research. Crisis informatics, or crisis computing, is an emerging interdisciplinary field, which combines computing and the knowledge of social science to extract disaster-related information (Soden & Palen, 2018). In this field, a major focus is to use and exploit social media data due to its timely availability and abundance. Social media analytics is the term that is commonly used for the analysis of social media data by combining and adopting different computational methods (Stieglitz, Mirbabaie, Fromm, & Melzer, 2018b; Stieglitz, Mirbabaie, Ross, & Neuberger, 2018d). Several studies highlight the benefits of social media analytics, which builds upon social media platforms such as Twitter and Facebook, for curating, analyzing and summarizing crisis-related information to help decision-makers and relief efforts (Imran et al., 2014, 2015; Nazer, Xue, Ji, & Liu, 2017; Reuter & Kaufhold, 2018; Terpstra, De Vries, Stronkman, & Paradies, 2012; Tsou et al., 2017; Vieweg, Hughes, Starbird, & Palen, 2010). The related literature in this domain can be divided into two categories, namely (i) "situational awareness," which corresponds to research work that focuses on understanding the big picture during a disaster event; (ii) "actionable," which corresponds to works that focus on identifying and extracting actionable information that enables first responders to help victims.

Situational Awareness

The concept of situational awareness refers to the understanding of the "big picture" in an emergency situation (Vieweg, 2012). The more situational awareness formal humanitarian organizations and people have, the better prepared they are to make informed decisions. Several studies contribute towards enhancing situational awareness of an event using social media, and most of them focus on Twitter mainly because of its timeliness and availability of information from a large user base. For instance, Kryvasheyeu et al. (2016) presented that physical disaster effects as well as real and perceived threats are directly observable through the intensity and composition of Twitter messages. Moreover, Hagen, Keller, Neely, DePaula, and Robert-Cooperman (2017) analyzed Twitter network structure to understand the flow of information and how different actors and communities contribute towards influential topics. Similarly, Kim and Hastak (2018) explored how emergency agencies and organizations can better plan operation strategies for a disaster by utilizing individuals' information on a social network.

Inspired by such studies, Landwehr, Wei, Kowalchuck, and Carley (2016) presented a Twitter tracking and analysis platform, called TWRsms, and elaborate on the associated challenges to deploy such a system. Their system targets a particular place, Padang Indonesia. They reported that using this system they can identify where the population is, who are the local opinion leaders, and also the content of the tweets. Their findings suggest that identifying local opinion leaders can be helpful for early warning. Later on, Avvenuti, Cresci, Del Vigna, and Tesconi (2017) proposed "Earthquake Alerts and Report System," which exploits tweets to understand how such a system can be useful during crisis-related events. The system collects tweets during an ongoing crisis event, filters irrelevant content, detects an event, assesses damage, and for the sake of comprehensibility, it provides a visualization. The authors concluded that such a system is highly important for crisis-related events.

Since there are overwhelming amount of social media data, finding relevant information for a target event is a challenging task. Several approaches have been proposed in (Laylavi, Rajabifard, & Kalantari, 2017; Robinson, Power, & Cameron, 2013; Sakaki, Okazaki, & Matsuo, 2010; Weiler, Grossniklaus, & Scholl, 2016; Weng & Lee, 2011) for event detection from Twitter streams. For instance, Laylavi et al. (2017) proposed a term-frequency based approach for detecting crisis-related events such as storm and/or bushfire whereas Sakaki et al. (2010) and Robinson et al. (2013) proposed similar approaches for detecting earthquakes from Twitter posts.

As for the crisis-related information classification, current literature broadly includes different typologies of messages by information provided (e.g., affected or injured people, infrastructure damage, urgent needs of affected population) (Aipe, Mukuntha, Ekbal, & Kurohashi, 2018; Alam, Joty, & Imran, 2018; Burel & Alani, 2018; Burel et al., 2017; Imran,

Elbassuoni, Castillo, Diaz, & Meier, 2013a, 2013b; Neppalli, Caragea, & Caragea, 2018; Nguyen, Al Mannai, et al., 2017a), or by information source (e.g., eyewitness accounts, official government sources, TV, radio) (Olteanu et al., 2014), or by information credibility factors (e.g., fake news, rumors, disinformation, misinformation) (Castillo, Mendoza, & Poblete, 2011; A. Hughes & Palen, 2012), or by temporal aspects (e.g., using different temporal phases of an event including pre, during, and post) (Chowdhury, Imran, Asghar, Amer-Yahia, & Castillo, 2013).

In addition to the textual content analysis, recent studies have also been focusing on the analysis of imagery content shared on social media (Alam, Imran, & Ofli, 2017; Alam, Ofli, & Imran, 2018b; Chen, Lu, Kan, & Cui, 2013; Daly & Thom, 2016; Li, Caragea, Caragea, Imran, & Ofli, 2019; Mouzannar, Rizk, & Awad, 2018; Nguyen, Alam, Ofli, & Imran, 2017b; Nguyen, Ofli, et al., 2017c; Peters & de Albuquerque, 2015). For example, Peters and de Albuquerque (2015) analyzed the data collected from Flickr and Instagram for the flood event in Saxony, 2013. Their findings suggest that the existence of images within on-topic textual content were more relevant to the disaster event, and the imagery content also provided important information, which is related to the event. The study by Daly and Thom (2016) analyzed images extracted from social media data, which is focused on a fire event. They analyzed spatio-temporal meta-data associated with the images and suggested that geotagged information is useful to locate the fire affected areas. The analysis of imagery content shared on social media has been explored using deep learning techniques in several studies to assess damage severity automatically (Alam et al., 2017; Nguyen, Alam, et al., 2017; Nguyen, Ofli, et al., 2017c). Nguyen, Ofli, et al. (2017c) used images shared on Twitter to assess the severity of infrastructure damage. Recently, Alam, Ofli, and Imran (2018b) presented an image processing pipeline to extract meaningful information from social media images during a crisis situation, which has been developed using deep learning-based techniques. Their image processing pipeline includes collecting images, removing duplicates, filtering irrelevant images, and finally classifying them with damage severity.

Combining textual and visual content can provide highly relevant information as discussed in (Alam, Ofli, & Imran, 2018a; Bica, Palen, & Bopp, 2017; Chen et al., 2013; Liang, Caverlee, & Mander, 2013). Bica et al. (2017) explored social media images posted during two major earthquakes in Nepal during April–May 2015. Their study includes identifying geo-tagged images and their associated damage. Chen et al. (2013)

studied the association between tweets and images, and their use in classifying visually relevant and irrelevant tweets. They designed classifiers by combining features from the text, images and socially relevant contextual features (e.g., posting time, follower ratio, the number of comments, re-tweets), and reported an F1-score of 70.5% in a binary classification task, which is 5.7% higher than the text-only classification. Mouzannar et al. (2018) explored damage detection by focusing on human and environmental damages. Their experimental methodology includes unimodal as well as multimodal analyses with decision- and feature-level fusion, and shared representations.

Actionable Insights

In contrast to the concept of situational awareness, actionable information research works focus on how to obtain the right information for the right person (i.e., disaster responder) at the right time during a disaster event. For instance, Zade et al. (2018) conducted a large-scale survey and interviewed several officials from humanitarian organizations. The authors identified that the notion of actionability differ from responder to responder, and there are several factors that inform whether a piece of information is actionable or not. These factors include timeliness, location, information source credibility, responder's role, and context.

Compared to situational awareness research, very limited focus has been given to learn the actionability of social media content during disasters. A recent study by Avvenuti, Cresci, Del Vigna, Fagni, and Tesconi (2018) investigates this crucial aspect and presents a system called CrisMap, which extracts potential crisis-related actionable information from tweets by adopting a classification technique based on word embeddings and by exploiting a combination of readily-available semantic annotators to geo-parse tweets. The system then visualizes the extracted information in customizable web-based dashboards and maps. This is very initial step towards modeling actionability of social media content, which needs more focus from the research community.

DATASETS

In this section, we present an overview of the publicly available social media datasets in the crisis informatics literature. These datasets are collected mostly from Twitter during disasters and emergencies. Most popular datasets include CrisisLex (Olteanu et al., 2014), CrisisNLP (Imran et al., 2016), and CrisisMMD (Alam, Ofli, & Imran, 2018a).

CrisisLex[1] comprises two different datasets, i.e., CrisisLexT26 and CrisisLexT6. The former dataset, i.e., CrisisLexT26, contains ~28 K tweets from 26 different crisis events that took place in 2012 and 2013 (Olteanu et al., 2014). This dataset was prepared to explore two different crisis dimensions. First dimension is the disaster type (natural vs. human-induced), their sub-type (e.g. meteorological, hydrological, etc.), temporal characteristics (instantaneous vs. progressive), and geographic spread (concentrated vs. diffused). The second dimension is the content type in which several types of categories and sub-categories were identified, including informativeness (informative vs. not-informative), information type (six different subcategories), and source of information (e.g., an eyewitness, government, NGOs, Business, Media, and Outsiders). The latter dataset, i.e., CrisisLexT6, consists of ~60 K tweets from six disasters that took place between October 2012 and July 2013 in USA, Canada, and Australia. Crisis keywords and locations have been used during the data collection process (Olteanu et al., 2014). Annotation task includes related vs. not-related messages. Both datasets have been annotated using Figure Eight[2] (previously known as CrowdFlower).

CrisisNLP,[3] on the other hand, consists of ~50 K tweets, which have been collected from 19 different disasters that took place between 2013 and 2015. The dataset was annotated in two steps. The first set of annotations was obtained using Stand- By-Task-Force (SBTF)[4] volunteers. The second set of annotations was curated using Figure Eight. The annotations consist of nine different humanitarian information types (e.g., injured or dead people; missing, trapped, or found people; displaced people and evacuations; infrastructure and utilities damage; donation needs or offers or volunteering services; caution and advice; sympathy and emotional support; other useful information; and not-related or irrelevant).

Apart from these prominent datasets, there are a few other, relatively small-scale datasets available online. Disaster Response Dataset[5] contains 30 K tweets collected during disasters such as an earthquake in Haiti in 2010; an earthquake in Chile in 2010; floods in Pakistan in 2010; Hurricane Sandy in USA in 2012, and news articles. The annotations include 36 different categories. Disasters on Social Media dataset[6] contains 10 K tweets collected and annotated with labels as related vs. not-related to the disasters. Social Web for Disaster Management (SWDM) dataset[7] consists of two data collections. The Joplin collection contains tweets collected during the tornado that struck Joplin, Missouri on May 22, 2011. The Sandy collection contains tweets collected during Hurricane Sandy, that hit Northeastern US on Oct 29, 2012. The Joplin dataset consists of 4400 labeled instances and the Sandy dataset consists of 2000 labeled instances (Imran et al., 2013b).

In contrast to vast availability of social media text datasets, there is scarcity of social media image datasets curated for disaster response and management. Damage Assessment from Social Media Imagery dataset[8] is a recent, first-of-its-kind example that aims to bridge this gap (Nguyen, Ofli, et al., 2017c). It consists of ~26 K images collected from Twitter during four natural disasters, namely Typhoon Ruby in 2014, Nepal Earthquake in 2015, Ecuador Earthquake in 2016, and Hurricane Matthew in 2016. This initial Twitter image collection was later supplemented by images collected from Google using queries such as "damage building," "damage bridge," and "damage road" to deal with labeled data scarcity problem. The final set of images were then annotated into three categories based on the severity of damage content (e.g., severe, mild, and little-to-none) using Figure Eight.

Taking a step further to maximize the information utility of social media data, CrisisMMD[9] paves the way to create grounds for multimodal analysis of textual and imagery content for disaster response and management (Alam, Ofli, & Imran, 2018a). This multimodal dataset consists of ~16 K tweet texts and ~18 K associated images with ground-truth annotations for several tasks. Tweets have been collected from seven natural disasters that took place in 2017. The annotations include three tasks: (i) informative vs. not-informative, (ii) humanitarian categories (eight classes), and (iii) damage severity (three classes). The third annotation task, i.e., damage severity, was applied only on images. The annotation for text and images was run independently for each event using Figure Eight.

Systems

In this section, we provide an overview of publicly available tools and systems that utilize social media data to extract crisis-related information.

Ushahidi[10] is a platform that started its journey back in 2008 as a free and open- source platform that allowed interested individuals and groups to create live, interactive maps (Meier, 2012; Okolloh, 2009). Initially, it aimed to visualize post-election messages on a map. Its current functionality includes mapping and visualization tools to create real-time, dynamic, and multifaceted crisis maps. It allows for visualizing how a crisis situation is evolving over time. A notable deployment of the system is the 2010 Haiti earthquake (Goolsby, 2010).

TweetTracker[11] is an online application aimed at supporting humanitarian aid and disaster relief responders to monitor and analyze social media content during crisis situations (Kumar, Barbier, Abbasi, & Liu, 2011). The functionality of the tool includes monitoring and analyzing location, real-time trends, and historical review of keyword-specific Twitter data.

Tweedr[12] is another tool that aimed to extract actionable information during natural disasters from Twitter. The tool employs clustering and classification techniques on tweet messages.

Twitris[13] application is also targeted for Twitter content analysis (Purohit & Sheth, 2013). The system is capable of collecting, aggregating, and analyzing tweets to provide deeper insights as well as to facilitate research and development process.

Artificial Intelligence for Disaster Response (AIDR)[14] is an active system that allows users (e.g., humanitarian organizations) to collect tweets during disasters and create custom classifiers on-the-fly by annotating incoming Twitter data according to the categories of their own choices (Imran et al., 2014). AIDR has been used my many humanitarian organizations including the United Nations Office for the Coordination of Humanitarian Affairs (OCHA) during many major disasters in the past.

SensePlace2[15] is an application focusing more on extracting spatio-temporal information with analytical capabilities to deal with large volumes of tweets and on providing visual interface methods to enable understanding of place, time, and theme components of evolving situations (MacEachren et al., 2011).

Enhanced Messaging for the Emergency Response Sector (EMERSE)[16] system has been developed to analyze tweets and SMS messages (Caragea et al., 2011). The major capabilities of this system include Twitter crawler, machine translation system, and classification modules while also supporting users with an iPhone application.

Emergency Situation Awareness (ESA)[17] is aimed at enhancing situational awareness with respect to crises induced by natural hazards, particularly earthquakes (Yin, Lampert, Cameron, Robinson, & Power, 2012). This tool focuses on spatio-temporal information extraction and its features include event detection, text classification, online clustering, and geotagging.

Twicident[18] is a framework and a Web-based system that automatically tracks and filters information relevant to real-world incidents and crises (Abel, Hauff, Houben, Stronkman, & Tao, 2012).

Pipeline for Post-Crisis Twitter Data Acquisition is an ongoing effort that aims to develop a system that requires minimal supervision in an active learning framework (Kejriwal & Gu, 2018).

Botometer[19] is an application aimed to detect accounts that are controlled by automated tools, termed as social bots (Varol, Ferrara, Davis, Menczer, & Flammini, 2017). Such an application is important to detect and filter fake or manipulated content on social media.

Crisis Event Extraction Service (CREES)[20] is an open-source web API to automatically classify social media posts during crisis situations (Burel & Alani, 2018). The system is developed for three different tasks, where the classification models are based on deep learning algorithms. One can deploy the system on their own server to use the API functionality. This system is currently integrated into the Ushahidi platform as part of the COMRADES project.[21]

Crisis Data Processing Services (CrisisDPS)[22] are designed to be simple, efficient, and accessible to both technical and non-technical end-users including crisis man- agers, humanitarian experts, first responders, and volunteering organizations, among others (Alam, Imran, & Ofli, 2019). Similar to CREES, CrisisDPS provides three types of data processing services, namely (i) disaster type classification, (ii) informativeness classification, and (iii) humanitarian information type classification. Each classification service offers three types of functionality: (i) single-item processing API, (ii) batch processing API, and (iii) file processing service.

Discussion and Conclusion

In this section, we discuss the limitations of the existing datasets, research studies, and tools and conclude with future directions and trends.

Although there are several small- to medium-scale public datasets in the literature, there is still a strong need for large-scale annotated datasets in multiple modalities (e.g., text, image, video, etc.), and preferably geotagged. For instance, in computer vision domain, one of the most important factors behind the rapid advancement of the field in the last decade or so is the increasing number of large-scale publicly available datasets for various domain-specific problems. Similarly, crisis informatics community should dedicate more time and effort to create large-scale annotated datasets to help flourish research and development activities in the domain.

On the other hand, collecting large amounts of annotated data for every single disaster event is not possible. The idea of using data from past disasters is a promising option, although sub-optimal, since using data from the current disaster usually yields better results. Therefore, methods such as domain adaptation and transfer learning can be employed to utilize data from past events, possibly in combination with unlabeled or few labeled data from the new event (i.e., zero-shot or few-shot learning). In addition, online and active learning techniques can be used to update the existing models upon receiving new training data samples to dynamically adapt to new changes and patterns in the data.

Gaining situational awareness is a core task of many humanitarian organizations. However, in many cases, they also need more actionable bits of information. These may be implicit and explicit requests related to emergency needs that should be fulfilled or serviced as soon as possible. Despite extensive research that mainly focuses on situational awareness from social media platforms, limited work has been done on understanding how actionable each piece of information is in a given context and for a given user. Computational techniques that can (i) automatically identify actionable messages from a live data stream during emergency events, (ii) assess their urgency, and (iii) categorize them according to information needs of humanitarian organizations are essential for crisis responders to launch rapid relief efforts.

Despite social media quickly becoming non-traditional data sources in many domains, social media platforms are usually noisy communication channels that can be polluted by misinformation (unintentional), disinformation (intentional), and rumors (unverifiable). Therefore, a major challenge is determining credibility of the information extracted from social media before including that bit of information in any critical decision-making process. Despite several recent attempts, information credibility issue remains largely unsolved.

On the systems and services side, despite the fact that extensive research has been conducted over the last decade to develop technologies to support humanitarian aid tasks, many of the developed technologies are still limited as they require both manual and automatic approaches, and more importantly, are not ready to be fully integrated into the disaster response workflows.

Notes

1. https://crisislex.org
2. https://www.figure-eight.com/
3. http://crisisnlp.qcri.org/
4. www.standbytaskforce.org
5. https://www.figure-eight.com/dataset/combined-disaster-response-data/
6. https://data.world/crowdflower/disasters-on-social-media
7. http://crisisnlp.qcri.org/
8. http://crisisnlp.qcri.org/
9. http://crisisnlp.qcri.org/
10. https://www.ushahidi.com/
11. http://tweettracker.fulton.asu.edu/
12. https://github.com/dssg/tweedr/
13. http://twitris.knoesis.org/
14. http://aidr.qcri.org/
15. https://www.geovista.psu.edu/SensePlace2/
16. http://emerse.ist.psu.edu/
17. https://esa.csiro.au/aus/index.html
18. http://www.wis.ewi.tudelft.nl/twitcident/
19. https://botometer.iuni.iu.edu/\#\#!/api
20. https://evhart.github.io/crees/
21. https://www.comrades-project.eu/
22. https://crisisdps.qcri.org

References

Abel, F., Hauff, C., Houben, G.-J., Stronkman, R., & Tao, K. (2012). Semantics + Filtering + Search = Twitcident. Exploring Information in Social Web Streams. In *Proceedings of the 23rd ACM HT Conference* (pp. 285–294).

Aipe, A., Mukuntha, N., Ekbal, A., & Kurohashi, S. (2018). Deep Learning Approach Towards Multi-Label Classification of Crisis Related Tweets. In *Proceedings of the 15th ISCRAM*, 2018.

Alam, F., Imran, M., & Ofli, F. (2017, August). Image 4act: Online Social Media Image Processing for Disaster Response. In *International Conference on Advances in Social Networks Analysis and Mining (ASONAM)* (pp. 1–4).

Alam, F., Imran, M., & Ofli, F. (2019, May). CrisisDPS: Crisis Data Processing Services. In *International Conference on Information Systems for Crisis Response and Management (ISCRAM)*.

Alam, F., Joty, S., & Imran, M. (2018). Domain Adaptation with Adversarial Training and Graph Embeddings. In *56th Annual Meeting of the Association for Computational Linguistics (ACL)*. ACL.

Alam, F., Ofli, F., & Imran, M. (2018a). CrisisMMD: Multimodal Twitter Datasets from Natural Disasters. In *Proceedings of International AAAI Conference on Web and Social Media (ICWSM)*.

Alam, F., Ofli, F., & Imran, M. (2018b). Processing Social Media Images by Combining Human and Machine Computing During Crises. *International Journal of Human Computer Interaction, 34*(4), 311–327.

Avvenuti, M., Cresci, S., Del Vigna, F., Fagni, T., & Tesconi, M. (2018). CrisMap: A Big Data Crisis Mapping System Based on Damage Detection and Geoparsing. *Information Systems Frontiers*, 1–19.

Avvenuti, M., Cresci, S., Del Vigna, F., & Tesconi, M. (2017). On the Need of Opening Up Crowdsourced Emergency Management Systems. *AI & SOCIETY*, 1–6.

Bica, M., Palen, L., & Bopp, C. (2017). Visual Representations of Disaster. In *CSCW* (pp. 1262–1276).

Burel, G., & Alani, H. (2018). Crisis Event Extraction Service (CREES)-Automatic Detection and Classification of Crisis-Related Content on Social Media. In *Proceedings of the 15th ISCRAM*, 2018.

Burel, G., Saif, H., Fernandez, M., & Alani, H. (2017, May). On Semantics and Deep Learning for Event Detection in Crisis Situations. In *Workshop on Semantic Deep Learning (SEMDEEP)*, at ESWC 2017.

Caragea, C., McNeese, N., Jaiswal, A., Traylor, G., Kim, H.-W., Mitra, P., … others. (2011). Classifying Text Messages for the Haiti Earthquake. In *Proceedings of the 8th ISCRAM*, 2011.

Castillo, C. (2016). *Big Crisis Data*. New York: Cambridge University Press.

Castillo, C., Mendoza, M., & Poblete, B. (2011). Information Credibility on Twitter. In *Proceedings of the 20th International Conference on World Wide Web* (pp. 675–684).

Chen, T., Lu, D., Kan, M.-Y., & Cui, P. (2013). Understanding and Classifying Image Tweets. In *ACM International Conference on Multimedia* (pp. 781–784).

Chowdhury, S. R., Imran, M., Asghar, M. R., Amer-Yahia, S., & Castillo, C. (2013). Tweet4act: Using Incident-Specific Profiles for Classifying Crisis-Related Messages. In *10th International Conference on Information Systems for Crisis Response and Management (ISCRAM)* (pp. 1–5).

Daly, S., & Thom, J. (2016). Mining and Classifying Image Posts on Social Media to Analyse Fires. In *13th International Conference on Information Systems for Crisis Response and Management (ISCRAM)* (pp. 1–14).

Enenkel, M., Saenz, S. M., Dookie, D., Braman, L., Obradovich, N., & Kryvasheyeu, Y. (2018). Social Media Data Analysis and Feedback for Advanced Disaster Risk Management. In *Social Web in Emergency and Disaster Management*.

Frej, W. (2018). *Hurricane Florence Flood Victims Turn to Social Media for Rescue*. HuffPost. Retrieved May 7, 2019, from https://www.huffpost.com/entry/hurricane -florence-flood-victims-social-median5b9b86c0e4b013b097799cd1

Goolsby, R. (2010). Social Media as Crisis Platform: The Future of Community Maps/Crisis Maps. *ACM Transactions on Intelligent Systems and Technology (TIST), 1*(1), 7.

Gralla, E., Goentzel, J., & Van de Walle, B. (2013). *Field-Based Decision Makers Information Needs*. Digital Humanitarian Network, Geneva.

Hagen, L., Keller, T., Neely, S., DePaula, N., & Robert-Cooperman, C. (2017). Crisis Communications in the Age of Social Media: A Network Analysis of Zika-Related Tweets. *Social Science Computer Review*. https://doi.org/10.1177/0894439317721985

Hiltz, S. R., Kushma, J., & Plotnick, L. (2014, May). Use of Social Media by U.S. Public Sector Emergency Managers: Barriers and Wish Lists. In *11th International Conference on Information Systems for Crisis Response and Management (ISCRAM)* (pp. 602–611).

Hiltz, S. R., & Plotnick, L. (2013). Dealing with Information Overload When Using Social Media for Emergency Management: Emerging Solutions. In *10th International Conference on Information Systems for Crisis Response and Management (ISCRAM)* (pp. 823–827).

Hughes, A., & Palen, L. (2012). The Evolving Role of the Public Information Officer: An Examination of Social Media in Emergency Management. *Journal of Homeland Security and Emergency Management, 9*(1).

Hughes, A. L., & Palen, L. (2009). Twitter Adoption and Use in Mass Convergence and Emergency Events. *International Journal of Emergency Management, 6*(3–4), 248–260.

Imran, M., Castillo, C., Diaz, F., & Vieweg, S. (2015). Processing Social Media Messages in Mass Emergency: A Survey. *ACM Computing Surveys, 47*(4), 67.

Imran, M., Castillo, C., Lucas, J., Meier, P., & Vieweg, S. (2014). AIDR: Artificial Intelligence for Disaster Response. In *ACM International Conference on World Wide Web* (pp. 159–162).

Imran, M., Elbassuoni, S. M., Castillo, C., Diaz, F., & Meier, P. (2013a). Extracting Information Nuggets from Disaster-Related Messages in Social Media. In *10th International Conference on Information Systems for Crisis Response and Management (ISCRAM)* (pp. 791–801).

Imran, M., Elbassuoni, S. M., Castillo, C., Diaz, F., & Meier, P. (2013b). Practical Extraction of Disaster-Relevant Information from Social Media. In *Proceedings of the 22nd WWW* (pp. 1021–1024).

Imran, M., Mitra, P., & Castillo, C. (2016, May). Twitter as a Lifeline: Human-Annotated Twitter Corpora for Nlp of Crisis-Related Messages. In *Proceedings of the Tenth International Conference on Language Resources and Evaluation (LREC 2016)*. Paris, France: European Language Resources Association (ELRA).

Kejriwal, M., & Gu, Y. (2018). A Pipeline for Post-Crisis Twitter Data Acquisition. arXiv preprint arXiv:1801.05881.

Kim, J., & Hastak, M. (2018). Social Network Analysis: Characteristics of Online Social Networks after a Disaster. *International Journal of Information Management, 38*(1), 86–96.

Kryvasheyeu, Y., Chen, H., Obradovich, N., Moro, E., Van Hentenryck, P., Fowler, J., et al. (2016). Rapid Assessment of Disaster Damage Using Social Media Activity. *Science Advances, 2*(3), e1500779.

Kumar, S., Barbier, G., Abbasi, M. A., & Liu, H. (2011). Tweettracker: An Analysis Tool for Humanitarian and Disaster Relief. In *Proceedings of the 5th ICWSM, 2011*. AAAI Press.

Landwehr, P. M., Wei, W., Kowalchuck, M., & Carley, K. M. (2016). Using Tweets to Support Disaster Planning, Warning and Response. *Safety Science, 90*, 33–47.

Laylavi, F., Rajabifard, A., & Kalantari, M. (2017). Event Relatedness Assessment of Twitter Messages for Emergency Response. *Information Processing & Management, 53*(1), 266–280.

Li, X., Caragea, D., Caragea, C., Imran, M., & Ofli, F. (2019). Identifying Disaster Damage Images Using a Domain Adaptation Approach. In *16th International Conference on Information Systems for Crisis Response and Management (ISCRAM)*.

Liang, Y., Caverlee, J., & Mander, J. (2013, May). Text Vs. Images: On the Viability of Social Media to Assess Earthquake Damage. In *International Conference on World Wide Web (WWW)* (pp. 1003–1006).

Ludwig, T., Reuter, C., Siebigteroth, T., & Pipek, V. (2015). Crowdmonitor: Mobile Crowd Sensing for Assessing Physical and Digital Activities of Citizens during Emergencies. In *Proceedings of the 33rd Annual ACM Conference on Human Factors in Computing Systems* (pp. 4083–4092).

MacEachren, A. M., Jaiswal, A., Robinson, A. C., Pezanowski, S., Savelyev, A., Mitra, P., … Blanford, J. (2011). Senseplace2: GeoTwitter Analytics Support for Situational Awareness. In *Proceedings of the IEEE Conference on Vast, 2011* (pp. 181–190).

MacMillan, D. (2017). In Irma, Emergency Responders' New Tools: Twitter and Facebook. *The Wall Street Journal*. Retrieved May 5, 2019, from https://www.wsj.com/articles/for-hurricane-irma-information-officials-post-on-social-media-1505149661

Meier, P. (2012). Crisis Mapping in Action: How Open Source Software and Global Volunteer Networks Are Changing the World, One Map at a Time. *Journal of Map & Geography Libraries, 8*(2), 89–100.

Mouzannar, H., Rizk, Y., & Awad, M. (2018). Damage Identification in Social Media Posts Using Multimodal Deep Learning. In (pp. 529–543).

Nazer, T. H., Xue, G., Ji, Y., & Liu, H. (2017). Intelligent Disaster Response Via Social Media Analysis a Survey. *ACM SIGKDD Explorations Newsletter, 19*(1), 46–59.

Neppalli, V. K., Caragea, C., & Caragea, D. (2018). Deep Neural Networks Versus naïve Bayes Classifiers for Identifying Informative Tweets During Disasters. In *Proceedings of the 15th ISCRAM*, 2018.

Nguyen, D. T., Al Mannai, K. A., Joty, S., Sajjad, H., Imran, M., & Mitra, P. (2017a). Robust Classification of Crisis-Related Data on Social Networks Using Convolutional Neural Networks. In *Eleventh International AAAI Conference on Web and Social Media (ICWSM)* (pp. 632–635).

Nguyen, D. T., Alam, F., Ofli, F., & Imran, M. (2017b, May). Automatic Image Filtering on Social Networks Using Deep Learning and Perceptual Hashing During Crises. In *14th International Conference on Information Systems for Crisis Response and Management (ISCRAM)* (pp. 499–511).

Nguyen, D. T., Ofli, F., Imran, M., & Mitra, P. (2017c, August). Damage Assessment from Social Media Imagery Data During Disasters. In *International Conference on Advances in Social Networks Analysis and Mining (ASONAM)* (pp. 1–8).

Okolloh, O. (2009). Ushahidi, or Testimony: Web 2.0 Tools for Crowdsourcing Crisis Information. *Participatory Learning and Action, 59*(1), 65–70.

Olteanu, A., Castillo, C., Diaz, F., & Vieweg, S. (2014). Crisislex: A Lexicon for Collecting and Filtering Microblogged Communications in Crises. In *ICWSM*.

Peters, R., & de Albuquerque, J. P. (2015). Investigating Images as Indicators for Relevant Social Media Messages in Disaster Management. In *12th International Conference on Information Systems for Crisis Response and Management (ISCRAM)*.

Purohit, H., Hampton, A., Bhatt, S., Shalin, V. L., Sheth, A. P., & Flach, J. M. (2014). Identifying Seekers and Suppliers in Social Media Communities to Support Crisis Coordination. *Computer Supported Cooperative Work (CSCW), 23*(4–6), 513–545.

Purohit, H., & Sheth, A. P. (2013). Twitris v3: From Citizen Sensing to Analysis, Coordination and Action. In *Proceedings of the 7th ICWSM*, 2013. AAAI Press.

Reuter, C., & Kaufhold, M.-A. (2018). Fifteen Years of Social Media in Emergencies: A Retrospective Review and Future Directions for Crisis Informatics. *Journal of Contingencies and Crisis Management, 26*(1), 41–57.

Rhodan, M. (2017). 'Please Send Help.' Hurricane Harvey Victims Turn to Twitter and Facebook. *Time*. Retrieved May 7, 2019, from http://time.com/4921961/hurricane-harvey-twitter-facebook-social-media/

Robinson, B., Power, R., & Cameron, M. (2013). A Sensitive Twitter Earthquake Detector. In *Proceedings of the 22nd International Conference on World Wide Web* (pp. 999–1002). New York: ACM. Retrieved from http://doi.acm.org/10.1145/2487788.2488101

Rudra, K., Banerjee, S., Ganguly, N., Goyal, P., Imran, M., & Mitra, P. (2016). Summarizing Situational Tweets in Crisis Scenario. In *ACM Conference on Hypertext and Social Media* (pp. 137–147).

Sakaki, T., Okazaki, M., & Matsuo, Y. (2010). Earthquake Shakes Twitter Users: Real-Time Event Detection by Social Sensors. In *Proceedings of the 19th International Conference on World Wide Web* (pp. 851–860). New York: ACM.

Soden, R., & Palen, L. (2018). Informating Crisis: Expanding Critical Perspectives in Crisis Informatics. *Proceedings of the ACM on Human-Computer Interaction, 2*(CSCW), 162.

Starbird, K., Palen, L., Hughes, A. L., & Vieweg, S. (2010). Chatter on the Red: What Hazards Threat Reveals about the Social Life of Microblogged Information. In *ACM Conference on Computer Supported Cooperative Work* (pp. 241–250).

Stieglitz, S., Bunker, D., Mirbabaie, M., & Ehnis, C. (2018a). Sense-Making in Social Media during Extreme Events. *Journal of Contingencies and Crisis Management, 26*(1), 4–15.

Stieglitz, S., Mirbabaie, M., Fromm, J., & Melzer, S. (2018b). The Adoption of Social Media Analytics for Crisis Management-Challenges and Opportunities. In *Proceedings of the 26th European Conference on Information Systems (ECIS)*. Portsmouth, UK.

Stieglitz, S., Mirbabaie, M., & Milde, M. (2018c). Social Positions and Collective Sense-Making in Crisis Communication. *International Journal of Human–Computer Interaction, 34*(4), 328–355.

Stieglitz, S., Mirbabaie, M., Ross, B., & Neuberger, C. (2018d). Social Media Analytics– Challenges in Topic Discovery, Data Collection, and Data Preparation. *International Journal of Information Management, 39*, 156–168.

Terpstra, T., De Vries, A., Stronkman, R., & Paradies, G. (2012). *Towards a Realtime Twitter Analysis During Crises for Operational Crisis Management*. Simon Fraser University Burnaby.

Tsou, M.-H., Jung, C.-T., Allen, C., Yang, J.-A., Han, S. Y., Spitzberg, B. H., & Dozier, J. (2017). Building a Real-Time Geo-Targeted Event Observation (Geo) Viewer for Disaster Management and Situation Awareness. In *International Cartographic Conference* (pp. 85–98).

Varol, O., Ferrara, E., Davis, C., Menczer, F., & Flammini, A. (2017). *Online Human-Bot Interactions: Detection, Estimation, and Characterization*. In *Proceedings of the 11th ICWSM*, 2018.

Vieweg, S. (2012). *Situational Awareness in Mass Emergency: A Behavioral and Linguistic Analysis of Microblogged Communications*. Unpublished doctoral dissertation, University of Colorado at Boulder.

Vieweg, S., Castillo, C., & Imran, M. (2014). Integrating Social Media Communications into the Rapid Assessment of Sudden Onset Disasters. In *International Conference on Social Informatics* (pp. 444–461).

Vieweg, S., Hughes, A. L., Starbird, K., & Palen, L. (2010). Microblogging During Two Natural Hazards Events: What Twitter May Contribute to Situational Awareness. In *Proceedings of the Sigchi Conference on Human Factors in Computing Systems* (pp. 1079–1088).

Villegas, C., Martinez, M., & Krause, M. (2018). Lessons from Harvey: Crisis Informatics for Urban Resilience. In *Rice University Kinder Institute for Urban Research.*

Weiler, A., Grossniklaus, M., & Scholl, M. H. (2016, December). An Evaluation of the Run-Time and Task-Based Performance of Event Detection Techniques for Twitter. *Information Systems, 62*(C), 207–219.

Weng, J., & Lee, B.-S. (2011). Event Detection in Twitter. In *International AAAI Conference on Web and Social Media.*

Yin, J., Lampert, A., Cameron, M., Robinson, B., & Power, R. (2012). Using Social Media to Enhance Emergency Situation Awareness. *IEEE Intelligent Systems, 27*(6), 52–59.

Zade, H., Shah, K., Rangarajan, V., Kshirsagar, P., Imran, M., & Starbird, K. (2018). From Situational Awareness to Actionability: Towards Improving the Utility of Social Media Data for Crisis Response. *Proceedings of the ACM on Human-Computer Interaction, 2*(CSCW), 195.

CHAPTER 6

'Internet of Things' Applications in Disaster Management

Malavika Singh

What Is AI and IoT and Why Do They Matter to the Government?

Artificial intelligence is a term given to the cognitive ability of a non-living entity to imitate natural intelligence found in humans and animals. For millions of years, there is a self-sustained process that makes human life possible on our planet. Our environment has never been more out of balance than it is today, and this is manifested by the increase in the number of natural disasters. These disasters are signs that our planet is struggling to cope with imbalances, created as a result of industrial and population growth, by unleashing its natural intelligence such as floods, earthquakes, landslides etc. It is in these situations that the application of artificial intelligence, also known as machine intelligence, can help us in salvaging human life and reduce infrastructural and financial losses. Without the machine intelligence, some of the rescue and recovery operations would be too risky, too slow, too labor intensive or too costly for humans to perform. Artificial intelligence combined with the Internet of Things (IoT)

M. Singh (✉)
Amazon Alexa (Machine Learning Platform Services), Boston, MA, USA

T. V. V. Kumar, K. Sud (eds.), *AI and Robotics in Disaster Studies*,
Disaster Research and Management Series on the Global South,
https://doi.org/10.1007/978-981-15-4291-6_6

can be leveraged to achieve cost savings and greater efficiency in public service during rescue operations. The IoT is network of devices equipped with sensors and software that collect and transmit data. The number of IoT-connected devices is expected to be around 30 billion by 2020.[1] This means six IoT devices per individual on average. This gives agencies across the globe the opportunity to dive into the data generated by these devices. There is no limit to the potential usage of this data which can vary from understanding and tracking individual preferences, to studying characteristics of a population group to identify patterns in behavior. While governments have made some investments in this area, the potential of the IoT, especially in the public service space is largely untapped.

What Are the Opportunities with IoT for Public Service?

Sensors are devices with the purpose of capturing data from their environment. Once they collect this data, they send it to the cloud where it may be processed to enable the device to perform a certain action. Sensors are the building blocks for the Internet of Things and are part of all devices on the network. Their application includes consumer devices, home automation, entertainment, sports equipment, transportation, utilities sector, medical devices among other things. The combination of better communication platforms, improved sensor designs and reduced cost of data transmission have fueled the growth of IoT in various areas especially personal devices. Wearable devices use sensor technology to provide consumers the data they desire like measuring health parameters like blood pressure, or altitude for movement and tracking assistance, or depth in swim watches. Data collecting sensors can be designed to survive extreme environmental conditions such as exposure to corrosive liquids or pressure. Home automation devices use sensors in various appliances such as entertainment systems, cooking tools and even HVAC equipment. They monitor usage patterns for different devices in the house and share that data with inhabitants which helps the users become more conscious of their energy consumption. This helps reduce energy costs, waste and conserve water. Sensors also enable the collection of data in various medical applications which has changed the way in which patients are treated and monitored. Patients can now be monitored and treated remotely depending on the type of medical condition.

Artificial intelligence is achieved by training machines to think and act like humans. There are a wide variety of applications some of which are described above, where organizations leverage large quantities of data—audio, text, image, video etc., labeled by humans to train their machine learning models. These models are designed to reduce human effort and friction in completing regular tasks. They can improve the relevance of content in search engines, recommend a list of songs or movies based on your taste, and remind you when to call your aunt or to order potatoes or visit the dentist. They can predict stock markets trends, or allow autonomous vehicles to drive in their lanes.

Alexa-enabled devices, i.e. devices that have the Alexa voice assistant built into them, offer various 'skills' to their users. There are over 70,000[2] Alexa skills as of February 2019. Amazon even launched a Skill Finder feature to allow users to learn about the different skills supported by the voice assistant. You can invoke the skill by saying 'Alexa, open Skill Finder' to your Alexa device. These skills help users accomplish tasks across multiple domains like finance, productivity, fitness, entertainment and travel among other. As a software developer interested in building technology that has a positive impact on the environment and local communities, one could take a short course and start building these skills that will reach Alexa users across the globe in different languages. There are multiple skills today related to emergency preparedness and disaster management. One such skill is called 'Disaster Management' and can be enabled for free on an Alexa device. It can help adults and children learn about the precautions and safety measures to take under man-made or natural disasters. Accessing the skill is as simple as saying 'Alexa, open Disaster Management' or 'Alexa, start Disaster Management' or 'Alexa, launch Disaster Management'. Another similar skill is 'Disaster Global' that provides information about the latest natural disasters happening all over the world and about disasters geographically nearest to the user. The user can just say, 'I want to know about the latest disasters near me' to the Alexa device. There are a host of other Alexa skills that can help in emergency. Alexa can help you keep in touch with friends and family during a disaster. People mark themselves as safe on Facebook during earthquakes and other natural disasters. Alexa has a skill called 'Quick Status for Facebook'[3] that allows you to update your Facebook status using a voice command. Another skill with both a free and a paid version called 'Ask My Buddy' creates contact lists, allowing you to reach out to one or more of your contacts who are within shouting distance of their Alexa device. Other similar skills are 'The Guardian Circle' and 'My SOS Family' that allow specifying emergency contacts. Information on first aid or non-emergency medical advice is available via multiple skills such as 'First Aid Advice', 'Virtual Nurse', 'Mayo Clinic First Aid' among others. Detailed weather information can be obtained using skills such as 'The Weather Network' or 'Weather Bot'. Skills like 'Quake Alert' and 'Hurricane Alerts' pull information from the U.S. Geological Survey and from the National Hurricane Center respectively in case of catastrophic natural events. Similar skills can give you information on emergency supplies, phone numbers for FEMA, American Red Cross or other government agencies and shelters.

High value investments by governments: In the public service domain, a few city and state governments have been able to achieve big wins with relatively small investments in IoT.

The Lower Colorado River Authority (LCRA) uses 270 sensors, called Hydromet, along 600 miles of river to alert authorities to possible flood conditions.[4] The LCRA was created to help Texas better manage the flood plain along the Lower Colorado River Basin, as well as to produce and distribute hydroelectric power to the cities in the area. Over the last few decades, using its network of sensors, LCRA has been able to monitor and report stream flows and other data, including temperature, rainfall and humidity, on a public website in near real time. LCRA recently received funding from the U.S. Department of Homeland Security to investigate better sensor technologies and software needed to disseminate information and alerts during a flood. LCRA is now looking to drive down cost per sensor from $25,000 to $50,000 per head to $200 to $2500 per head while maintaining functions that help the sensors survive tough weather conditions. The goal is to build a better framework so that during an emergency, responders can send more targeted warnings to the smartphones of citizens in areas where flooding is imminent.

The city and the California Institute of Technology (Caltech) are doing the same thing using a project called 'Quake Alert', which uses sensors to detect and visually depict tremors in real time in the area. The city is developing a system of sending alerts to citizens' smartphones to give them 15–30 seconds to take cover.

The city of Chicago[5] uses a network of surveillance tools, biological, chemical and radiological sensors that feed data continuously to its operations center from which emergency response can be deployed.

The Rio de Janeiro[6] City Hall Operations Center uses IoT technology to manage weather, traffic, police and medical services in the city. After Hurricane Harvey, the city of Houston used IoT to identify damage and collect information.

Through the 'Internet of Trees' project, the city of Los Angeles is combining data from Google Street View with a machine learning algorithm developed by Caltech to inventory its urban forest of ~700,000 trees scattered over 469 square miles. This saved the city ~$3 million, which would have to be spent on people to physically inspect each tree.

Overall, machine learning models can do anything as long as you feed them the data. Then why is it so difficult to predict every natural disaster across the globe when 66.6%[7] of the world's population carries a cell phone- or tablet- or cellular-enabled IoT device? Why in spite of the availability of such sophisticated voice-controlled technical solutions do we not think of all such Alexa skills during a disaster?

Why Are Government Agencies Not Able to Leverage IoT?

In spite of the wide applications and all the opportunities associated with the IoT, governments across the globe have not invested enough in IoT projects. Lack of leadership, infrastructure, technical skills, lack of funding to modernize IT, issues of security, privacy, interoperability make leaders question the return on IoT investments. Let's look at some of these reasons and barriers that contribute to IoT remaining an academic interest.

Governments, especially those in developing nations like India, do not have a strong infrastructure to build on top of. This could be both technical and social infrastructure. Pre-existing legacy software systems, basic utilities like power and Internet, or social infrastructure like a literate population that could interpret text message warnings, complete census data, local laws, or other cultural norms could pose to be a barrier to entry for any new IoT initiative. The challenge remains to first identify the gaps in the existing processes and streamlining them, with or without technology. Building a technical solution on top of a broken system is likely to fail due to misalignment with local processes and lack of community support.

Today, less than 10% of organizations are able to capture and analyze IoT data timely. Most claim that faster analytics would help them to obtain results they seek from their IoT investments. Current practices focus on data collection and storage for future use; however, if governments can streamline processes and setup technology to analyze data and present insight in real time, the applications for public service and disaster management will increase significantly.

Very often, governments find themselves gathering data without setting clear guidelines on the problem to solve. As sensors and related technologies becomes cheaper, it is tempting to use them everywhere; however, without a clear purpose for the technology, governments will spend taxpayer money on collecting the data, setting up data centers, servers, support staff and maintenance without a purpose. Defining a purpose for the data is essential and so investing in data science and analytics should be a top priority for any public service organization. This investment early on ensures the investment in technology is aligned with the government's goal to innovate with data in the long run. Partnering with local universities, schools and colleges would be key to defining local problems correctly and ensuring the right skill set is available to solve the challenge—both academically and practically.

Another issue that makes governments risk averse is that related to security and privacy. We have not yet figured out a way to deal with a situation where IoT devices begin generating data incorrectly. For example, if a sensor in a car sends incorrect signals to the insurance company, will the driver of the car pay a higher insurance premium for exceeding speeds, or will it be the sensor manufacturer? We don't know. Similarly, if the government launches a large technical initiative to gather data from a million mobile subscribes, and there is a data leak on the network, the damage would far exceed the damage to individual privacy. The government may need to deal with both legal and financial issues, in addition to the loss of trust from the public in its policies and execution.

Phil Bertolini,[8] CIO of Oakland County, Michigan, expresses that he is extremely cautious when it comes to IoT. While central management and control of systems via the Internet can reduce costs and increase efficiency, it can also increase the danger that such systems can be hacked. 'What if a hacker shuts down the air conditioning in a data center, causing millions of dollars' worth of damage to computer equipment? Worse yet, a nefarious actor might take control of all those Internet-connected lights and plunge a city's entire downtown into darkness. As government, we have to be extra careful.'

What Should Governments Do?

Start small: Given that governments are dealing with taxpayer money and responsible for ensuring its judicious use, the caution being exercised by agencies is justified. To ensure governments continue innovating and do not block potential improvements to existing processes by being over cautious, small investments in smaller jurisdictions are recommended rather than large investments that come with higher risk. Once a project has been proven to produce results at a local level, it can be replicated in metropolises using a phased approach with interim milestones. In this context it is important to mention the introduction of Reliance subsidiary, Jio telecom provider in India which revolutionized the ICT based communication system in 2016.It demonstrated immensely significant services in disaster affected areas offering free voice calls forever and 1 GB of mobile data for just Rs.50 (USD 0.75 in 2016). It spread to 100 million subscribers within six months. Jio found that communication may still not be as expected to find problem resolutions in a situation as 500 million Indians didn't have smartphones. So a lightweight low-priced smartphone was introduced in 2017 which offered 4G services with a bare-bone camera to complete a holistic information dissemination. This is exactly what one needs in a country aspiring to have a robust disaster response mechanism and a decentralized resilience management system.

Prioritize projects that generate real-time insights and consequently generate more value for citizens and the government. Governments should attempt to create an information value loop where investments within IoT directly translate into improved actions by the government. Today, emergency response is dependent on incomplete information and this impacts how quickly a respondent can be notified. In 2011, only 15 % of Los Angeles 911 dispatchers successfully alerted Los Angeles Fire Department response units within the targeted 60-second timeframe. Scenarios such as these can be minimized with investments in technology that facilitates real-time action. Environment sensors are one example of such technology. They can register and report early indicators of an emergency or crime. Devices such as ShotSpotter can detect the sound of a gunshot and pinpoint its location within 10 feet. The device automatically alerts the police and speeds up the reaction time, while reducing reliance on witnesses. In India's rural Karnataka where a small group of workers manned telephone helplines of the Bengaluru Electricity Supply Company (BESCOM) has turned to AI powered systems to service over 9000 complaints it receives on an average in a day on its helplines. Further to facilitate work in this direction BESCOM is collaborating with Medical Intelligence and Language Engineering (or MILE) Lab at the Indian Institute of Science (IISc) to create an AI based complaint response system that can cater to 500 calls at a time in contrast to 60 in the existing system. This AI based system would overcome barriers related to language, pronunciation and other linguistic expressions while at the same time every complaint would get recorded with specific instructions.

Crowdsource data – social media: Many times governments can bypass infrastructural investments by developing partnerships to crowdsource data. Social media networks like Facebook, WhatsApp, Twitter etc. generate data where users voluntarily share information about themselves such as their location. Google released its 'Person Finder' app after the 2010 Haiti earthquake, and Facebook allows users to mark themselves 'safe' during a disaster. FEMA is now integrating Facebook[9] and Twitter into its emergency communications. Meanwhile the federal Office of Management and Budget[10] is working with FEMA to explore how to crowdsource information during emergencies. Emergency systems could also integrate this precise location data with local video and social media to give responders context well before they arrive at the scene. The police department of one major US city is already experimenting with combining video and social media with facial-recognition or social-network analysis software to help officers better investigate crimes and identify suspects.[11]

Risks Related to IoT Expansion

Risks related to IoT expansion primarily fall into two categories—health risks and privacy risks.

Technologies like sensors, 5G, Wi-Fi form the backbone of IoT and Smart Cities; however, governments today do not have many long-term studies highlighting the health risks of long-term exposure to these IoT building blocks. The primary reason is that these technologies have not been around long enough, and 10-year studies conducted on rodents are inconclusive given the test conditions. Also, for the limited studies conducted, it is important to note the outcome as well as the sponsor for the study before placing our trust in them. The State of Louisiana recently ordered a study on the harmful effects of 5G. The state felt the need to call upon an independent study stating that 5G may pose risks to the environment due to increased radio-frequency radiation. They also noted the possibility of a wide range of effects and that insurance companies may have placed exclusions in policies to exempt damage caused by 5G technology. The debate is still ongoing, but it is clear that governments need to proceed with caution on the expansion of 5G towers.

Social media is a click away for anyone with Wi-Fi or mobile phone access today. There are a larger number of studies conducted on Wi-Fi compared to 5G. Multiple studies on Wi-Fi claim that it causes oxidative stress, sperm/testicular damage, neuropsychiatric effects including EEG changes, apoptosis, cellular DNA damage, endocrine changes and calcium overload. These effects can be observed by other similar EMF exposures as well.

The other category of risks is related to individual privacy. While the network of IoT devices makes our lives simpler in many ways, it comes at the cost of signing away a little bit of our privacy. A data leak in the network can become a security and financial threat for the concerned individual especially in cases where personally identifying information like the SSN may be compromised. Mischievous elements may be able to hack into devices on the network for a home automation system and gather details about a family's routine posing a security risk to all members of the family. This was recently seen in the US when a criminal was able to hack into the baby monitor in a house.

Summary

AI and IoT are great tools with immense potential for application in the public service space, including natural and man-made disasters. They do however come with some risks, as well as some investments. Governments across the globe should continue to tread with caution in this area and implement IoT projects with clearly analyzed data and projected outcomes. By exercising caution in three main areas of (1) public health, (2) end user privacy and (3) IT spending, the government can lay out plans for implementing smaller projects and then scaling them where there is a strong value seen for the public.

Notes

1. https://gcn.com/articles/2018/11/26/iot-enabled-government.aspx?m=1 and https://spectrum.ieee.org/tech-talk/telecom/internet/popular-internet-of-things-forecast-of-50-billion-devices-by-2020-is-outdated
2. https://www.cnet.com/how-to/amazon-echo-most-useful-alexa-skills/ accessed 4th November 2019
3. https://www.gearbrain.com/amazon-alexa-skills-emergency-plan-2496956754.html accessed 4th November 2019.
4. https://gcn.com/articles/2017/09/20/iot-flood-sensors.aspx accessed 4th November 2019.
5. https://www.chicago.gov/city/en/depts/oem/provdrs/tech.html, accessed 1st November 2019.
6. https://10innovations.alumniportal.com/internet-of-things/iot-in-disaster-management-saving-lives-with-early-warning.html, accessed 4th November 2019.
7. According to GSMA real-time intelligence data, there are now over **5.15 Billion people with mobile devices worldwide**, – This means that **66.60% of the world's population** have a mobile device (cell phone, tablet or cellular enabled IoT devices.) - https://www.bankmycell.com/blog/how-many-phones-are-in-the-world accessed 4th November 2019.
8. https://www.govtech.com/The-Smartest-Thing-Phil-Bertolini-Ever-Did.html accessed 9th November 2019.
9. https://www.fema.gov/news-release/2017/10/10/social-media-provides-online-information-and-resources-survivors accessed 10th November 2019.
10. https://www.fedscoop.com/emerging-technology-omb-support-first-movers/ accessed 10th November 2019.
11. https://www2.deloitte.com/us/en/insights/focus/internet-of-things/iot-in-government.html#endnote-24, accessed 21st November 2019.

CHAPTER 7

Samvad: Reaching Out Through Radio and Wireless Network

Sampark

Disasters and Information Dissemination

Natural disaster can be one of the greatest obstacles to sustainable development and social security. There are certain regions and localities, which are recurrently exposed to hazards like floods or earthquakes. In this context mainstreaming disaster risk mitigation programme is an important task as it will be of help in protecting the development gains. The case study, Commencement of Change' highlights how Kolhapur, which is exposed to recurrent events of flash floods, has institutionalized mechanisms through planned interventions by introducing structural and non-structural changes to improve response coordination during disasters and create a new culture of disaster preparedness and mitigation amongst the communities. It was due to this effort, the district could avert a major catastrophe in 2018. The case depicts how most of the Disaster Risk Management (DRM) related tasks require appropriate decentralization of functions, devolution of authority and community participation to complement the centralized system. A strong case is sought to be made

Sampark (✉)
DDMA, District Magistrate Office, Kolhapur, India

T. V. V. Kumar, K. Sud (eds.), *AI and Robotics in Disaster Studies*,
Disaster Research and Management Series on the Global South,
https://doi.org/10.1007/978-981-15-4291-6_7

out for the need to strengthen capacities of the District Administration as well as local communities to develop sustainability of the DRM programme in the region by ensuring deployment of adequate financial, technical and manpower resources.

Situation in District in Previous Floods

Kolhapur, an affluent district of Maharashtra, which has always been at the forefront because of its advancement in agriculture, industries, art and cooperative movement, was worst affected by unprecedented floods in the years 2005, 2006, 2007, 2008 and 2016. The rains, accompanied by the sudden release of dam water as well as the accumulated backwater effect from dams, barrages and weirs resulted in the flooding of vast areas. Approximately 30,080 families in 992 villages and six cities of Kolhapur were gravely affected by this disaster. People fled to safety in order to save their lives. Transport and communication was severely disrupted and Kolhapur was disconnected from other districts, thereby creating further impediments in rescue and relief operations. On 30th July, flooding from the eleven dams of Kolhapur worsened the situation. There was a colossal loss of human life, livestock, property and revenue in year 2005.

Response of the District Administration

Lack of preparedness on part of the District Administration as well as the local communities was the key reason for the huge loss and damage that had occurred. Subsequently, the District Administration geared up and decided to take proactive steps to handle future catastrophes that may strike Kolhapur. It further reviewed the situation to map immediate priorities by organizing consultative meetings of government functionaries, non-governmental organizations (NGOs) and local communities at the district, taluka and village level.

Implementation of the Seven-Point Disaster Risk Management Programme

In order to effectively handle the crisis situation evolving through disasters, the District Administration, Kolhapur developed a seven-point agenda under the Maharashtra Disaster Risk Management (MDRM)

Programme. The entire thrust of the programme was on introducing non-structural changes for developing **(A)** capacity, **(B)** communication and **(C)** coordination of the local communities as well as the administration.

(A) Dissemination of Information

Special emphasis was given on information, education and communication (IEC) campaign to educate people about the DRM Programme and the do's and don'ts during disasters.

(B) Decentralization of Disaster Management System

In order to ensure a synchronized effort towards handling disasters within the district, the District Administration developed a three-tiered coordination mechanism. A Disaster Management Committee (DMC) was set up at the district level and twelve committees at the taluka level were constituted. Similarly DMCs were constituted at municipal council level and village level.

According to Mr. Sanjay Shinde, RDC Kolhapur, effective coordination between these three tiers resulted in detailed planning and speedy execution of relief and rescue operations before and during the 2018 flood. For instance, the village DMC undertook the task of micro-planning for each and every village in the pre-monsoon months. Resource mapping was done by the village DMC and an inventory of resources comprising of alternative temporary shelter for affected people and cattle, along with adequate food, water, milk, medication and fodder was developed and maintained by the disaster management teams in the village. The detailed planning also involved distribution of mosquito nets and coils for every family in the temporary shelters.

During the flood 2018, the village functionaries like Talhati, Police Patil, Medical Officer and other key persons stayed put in the affected villages during disaster and in collaboration with the five disaster management teams, arranged for prompt relief and rescue operations. To ensure provision of prompt relief and dissemination of accurate information regarding emergency situation to the aggrieved villagers, the village DMC, as per its Standard Operation Procedures (SOPs), coordinated extensively with other line departments at taluka and district DMCs such as Police, Irrigation, Agriculture, Health, Animal Husbandry, Water Supply and NGOs such as 'Jeevan Mukti Sanstha' Jivan Jyoti Seva sanstha, Adhaar

Rescue force, Pass Rescue force, Wajir Rescue force. The District Collector, Superintendent of Police and Chief Executive Officer of District Administration regularly inspected the flood-affected villages, supervised the relief and rescue operation and ensured provision and distribution of all requisite supports to the flood-affected villagers. These initiatives not only arrested the magnanimous impact of disasters, but also strengthened the trust of distraught villagers, who witnessed these functionaries' active and extended involvement in relief and rescue operations.

(C) **Preparation of Standard Operating Procedures**

In order to ensure effective interdepartmental vertical and horizontal coordination SOPs were prepared before the onset of the 2018 monsoon. All the line departments like Police, PWD, Fire, Irrigation, Agriculture, Animal Husbandry, Women and Child Development and Health were instructed to prepare their own SOPs, which described the roles and responsibilities of the each department during disasters. Periodic training programmes were also conducted to facilitate the implementation of SOP. This initiative of the District Administration provided the much-required clarity for the departments regarding execution of their duties during disasters and facilitated establishment of accountability within the administration.

(D) **Preparation of Village Development Plan**

The DRM office at Kolhapur, undertook on a war footing, the IEC Campaign throughout the district. In January 2018, participatory risk appraisal process was conducted in 1226 villages for eliciting active involvement and participation of the people to design 'Village Disaster Management Plan'. In each village, as part of the Disaster Management Committee, early warning, first aid, search and rescue, food and shelter management and water and sanitation teams were formed. Each team consisted of at least six members. Members were selected on the basis of their suitability in terms of age, knowledge and capability.

(E) **Provision of Training**

In the subsequent months the District Administration of Kolhapur sought to coordinate activities of the newly formed teams with the Talathis of the villages and the line departments like PWD, Irrigation, Agriculture and Police. This enabled identification of roles and responsibilities, as well as areas for capacity building to cope with future disasters. Accordingly, theoretical and practical training in the areas of first aid and search and rescue operations were provided to the teams.

(F) **Procurement of Equipment**

In order to equip the district with first aid, search and rescue operations, the State Government provided five inflatable motor boats. The District Administration also procured rescue equipment like 06 inflatable motor boats, life jackets, life buoys, etc. from MDRM programme funds. These equipment were kept at the disposal of the Administration and provided to the affected villages during rescue and relief operations.

(G) **Ameliorating Communication Channels**

Poor communication was one of the major lacunae in the flood management of previous floods. In order to rectify this, the District Administration paid specific attention to the telecommunication channels and took following measures to facilitate their smooth operations:

The devastating floods of 2005 had paralyzed the Kolhapur District. The tragedy went beyond numbers as everywhere, marooned people were left to fend for themselves. In many places, the administration could not respond promptly to the series of calamities that unfolded as communication links had broken down. The telephone lines were down. Radio was the sole instrument of connection left with the outside world, but that too was unable to provide any substantial information. Many villagers from Hathkanangale and Shirol Talukas were not given early flood warning due to which they were stranded in the villages. Mr. Anil Kore, Programme Executive, All India Radio (AIR), Kolhapur on 25th July 2005 was shunted between the District Collector's office and Police Control Room for gathering flood and rescue related information for the purpose of broadcasting the same. Considering the fact that radio was the only media available for disseminating flood related information at the given time, the exigency to acquire this information from appropriate authorities was high. It took Mr. Kore nearly three hours to collect information and broadcast the same.

Subsequently, the AIR team solicited contact details of Irrigation, PWD, Agriculture, Revenue, Health and Police Department and latest information about the flood and rescue updates from the District Collector's office.

- **Immediate Assistance**

Between 26th July to 9th August 2005, the AIR team visited nearby villages in Kolhapur to gather first-hand information from the people and broadcast the same. Due to non-availability of transport facility they were unable to visit severely impacted places. The team managed to visit one of the rehabilitation centres in the school premises in Narsobavadi, Shirol Taluka and interacted with around hundred refugees. The refugees presented their testimonies on the air and appealed for assistance. In response to the broadcast, help from organizations like Mayur Dudh Sangh, Gokul Dudh Sangh, Varana Dudh Sangh & Walalwalkar Trust poured in, in form of milk, food, clothes and medicines. Under the compelling circumstances, radio became an essential gateway to the rest of the world, making it possible to allocate assistance and donations to those affected by the disaster. The AIR network clearly upheld the tradition of human solidarity.

Throughout the monsoon, the radio served as a link between the authorities and the population. It relayed the authorities' instructions to people, provided information on do's and don'ts of the flood, early warnings and traffic related information. This information was of high utility value both to the people as well as the government.

- **Preparedness Education: Use of Radio**

Communication plays a pivotal role in mobilizing people through advocacy, making people aware about the nature and content of any preparedness programme and enables them to understand in proper perspective the impact such a preparedness process would have on their lives, their households and communities. In this context, post disaster, in September 2005, **Mr. Prasad Sankpal**, (DPO) requested the Masters of Social Work (MSW) students from D. K. Shinde College, Shahu Institute to approach AIR, Kolhapur to consider the possibility of using radio services for generating awareness programmes and community involvement in 'Disaster Risk Management' (DRM). Considering the importance of the subject AIR readily supported the proposal. Accordingly, two special sessions with

Mr. Prasad Sankpal in October 2005 on 'Disaster Management' were incorporated in the 'Phone In' programme. People from all over the district contacted him and sought his advice on involving themselves in the DRM activities. Post the programme, the President of an NGO from Kagal Taluka approached Mr. Sankpal and expressed his wish to be involved in DRM activities. He was requested to register himself with the Disaster Management Committee of Kagal Taluka.

- **Initiatives Taken in 2006**

In order to disseminate information about disaster preparedness and mitigation, from January to March 2006, AIR conducted thirteen programmes of fifteen minutes each, every Sunday at 8.15 am to provide basic information on (a) disaster management (b) do's and don'ts of each disaster (fire, flood, earthquake, land sliding and cyclone) (c) information on Disaster Management Committee at the District level, Taluka level and Village level and (**D**) important emergency contact numbers.

To generate accurate information at the time of disasters, a control room was set up at the District level. Further, Deputy Collector formed 'Information Management Committee' comprising of Heads of Irrigation, PWD, Agriculture, Revenue, Health and the Police Department. The Committee members were informed to be present in the control room for 24 hours at the time of any disaster and continuously upgrade the information. This system helped AIR to get quick and accurate information. It also helped in winning people's confidence in the administration. In order to ensure that the system functioned, AIR warned the government authorities that if they do not provide necessary information at the time of disaster, they would broadcast the same.

During the 2006 monsoon, AIR relayed weather information, traffic information, and information about rains and floods at the interval of every half an hour. During floods the information was upgraded by every ten minutes. To build credibility in the news, messages and interviews of District Officials were also broadcast. This helped people in getting the right kind of information by contacting the right sources through the given contact numbers. The Deputy Collector personally requested people to refrain from believing any kind of rumours. The AIR team toured the flood-affected areas and directly connected people and government authorities through air. This facilitated the response process.

Impact of Radio Initiatives

The proactive initiatives taken by AIR helped in disseminating disaster management awareness, instructions for preventing disasters and information concerning behavioural conduct at the times of disasters. Early disaster warnings through radio helped people to take appropriate actions at the time of disaster and shift themselves and their animals to safer places. Since radio news was frequently updated and messages were directly relayed by the Deputy Collector and DPO, DRM, there was marginal scope for spreading rumours. Direct access with government officials through air provided immediate mobilization of assistance through public and private support.

Preparedness Education: Use of Cinema Halls

Cinemas are the most popular form of popular entertainment amongst the youth in small cities and towns. The use of cinemas for facilitating attitudinal changes required for developmental activities is not new. But the use of cinema hall for projecting disaster management education is definitely innovative. To promote the disaster preparedness campaign further, **Mr. Prasad Sankpal**, in October 2005, approached **Mr. Prashant Dhage**, Officer In Charge, Entertainment Department with the proposal to project the 'Do's & Don'ts messages during disasters like fire, earthquake, flood and landslides, in the cinema halls in fifteen theatres of Kolhapur. (See Sects. 7.1 and 7.2 for sample messages) Considering the fact that each theatre projects three shows per day, fifteen theatres of Kolhapur can disseminate disaster management information to more than 10,000-heterogeneous viewers. In the context of its relevance and importance of the subject the proposal was immediately accepted. The messages through slides have been projected in all the shows across all the fifteen theatres since March 2006 to September 2006 and March 2007 to September 2007.

Mr. Survanshi, Entertainment Tax Inspector (ETI) personally visits the cinema halls to monitor the regularity and consistency of projecting the disaster preparedness campaign. As a result of this information dissemination strategy, the District Administration received acknowledgements and appreciation from the public both through telephone calls and personal visits.

To evaluate the impact of these messages on the audience, a feedback survey was undertaken by the MSW students of **D. K. Shinde** College Shahu Institute, Kolhapur, in November 2005. The survey revealed that the audience found the disaster preparedness information very useful and helped them to gain knowledge. In order to percolate the information further they felt that the messages have to be projected for longer duration on regular basis.

Future Plan

As part of its future strategy AIR, Kolhapur plans to collaborate and partner with other media for developing more innovative programmes. It plans to disseminate disaster management information through folk songs like Powada and street plays. This medium will help in retaining information. To inspire villagers to take a proactive stand, it plans to broadcast village level meetings for disaster management.

The Final Word

Using the social marketing approach to promote the disaster preparedness theme, the DRM office of Kolhapur was able to plug into a commercially successful network of **'Radio and Cinema Hall'** that assured them of reaching a large, attentive audience for a very low cost. The bedrock of the campaign was the principle that people in a relaxed mode are more receptive to absorb information, which can facilitate behavioural intention and changes, which are a prerequisite for building up disaster management culture.

CHAPTER 8

Usages of AI Technologies in Nepal's Disaster Management

Gajendra Sharma and Subarna Shakya

Introduction

AI tries to improve and enhance the efficiency of the disaster management process. AI devices such as robotics are helpful to search and rescue operations, enhancing information sharing using ontologies, providing information to crisis factors and providing multi-agent systems for real-time support and simulated environments (Massaguer, Balasubramanian, Mehrotra, & Venkatasubramanian, 2006). The success of emergency response depends on gathering information from distributed sources, integrating them and making appropriate decisions. Such complexity

G. Sharma (✉)
Department of Computer Science & Engineering, Kathmandu University, Dhulikhel, Nepal
e-mail: gajendra.sharma@ku.edu.np

S. Shakya
Institute of Engineering, Pulchowk Campus, Tribhuvan University, Lalitpur, Nepal
e-mail: drss@ioe.edu.np

T. V. V. Kumar, K. Sud (eds.), *AI and Robotics in Disaster Studies*, Disaster Research and Management Series on the Global South,
https://doi.org/10.1007/978-981-15-4291-6_8

makes impossible for any single individual or team of people to fulfill the roles sufficiently. Ontologies and semantic web are adopted to solve integrating problems, for example ontologies are used in integrating heterogeneous information sources and semantic web services are used to provide customized queries to crisis. The World Wide Web Consortium (W3C) and e-response represent the visible effort in building disaster response ontologies and getting the benefits of semantic web services (Disaster Management, 2019).

Nepal in this case is prone to disaster issues every year and faces a number of challenges. Figure 8.1 shows the geography of Nepal. It is divided into three major parts: Terai (plain area), hilly region and Himalayan region. The major disasters frequently happen in Nepal are earthquakes, landslides, flooding and avalanches and glacial lake outburst. Nepal earthquake killed nearly 9000 people and injured nearly 22,000 at 11:56 on 25 April 2015, with a magnitude of 8.1Ms (Fig. 8.2).

In disaster management a multi-agent system provides the decisive solution to entire problems related to interaction and coordination of response teams. Multi-agent systems for crisis response include real-time support and simulation systems such as DrillSim, DEFACTO and WIPER. Multi-robot solutions had been used in a wide range of crisis response. Usually, robots are used in Urban Search and Rescue (USAR) operations. USAR includes locating, rescuing and stabilizing victims

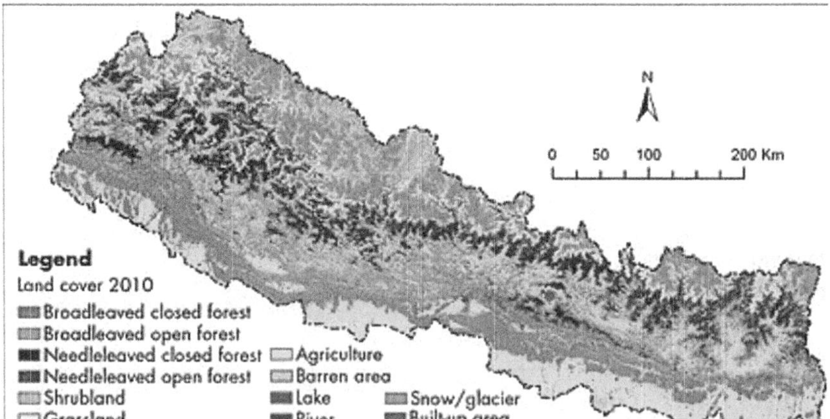

Fig. 8.1 Geography of Nepal (Geography of Nepal, 2020)

Fig. 8.2 Nepal earthquakes of 2015

trapped in trapped places. USAR workers have 48 hours to find trapped survivors in a collapsed structure otherwise the probability of finding victims is not possible. Greer (2002) found challenges that USAR team have to overcome into four major areas: (1) efficient response (2) rescuers safety (3) environment disturbance and climatic conditions and (4) inappropriate equipment and resources.

Information management and processing in emergency reaction are responsible to produce digital representations for common response operations. This cannot be effective without overcoming the following challenges (Mehrotra et al., 2004): (1) Variety of information sources such as information relevant to decision making is dispersed from sensors where data is created to heterogeneous databases belonging to autonomous organizations. A significant information spans different modalities such as voice conversations among crisis responders, cameras data, sensor data, geographical information systems data and relational information in databases (2) Diversity of information users: different people or organizations have diverse needs and urgency levels regarding the same information. According to these challenges different types of data are used, but a common set is shared all over. This common set of information can be represented by ontology.

A multi-agent system (MAS) is composed of multiple interacting intelligent software systems. They can be used to solve problems which are difficult for individual to solve such as disaster response, and formulating social structures. In the meantime, multi-agent architecture is the spirit of response systems. The original concept arises from agent characteristics in MAS, such as independence, local view of environment, and capability of planning, learning, coordination and decision making. Bearing in mind the amount of information that Twitter has, it is challenging for emergency managers and stakeholders to investigate individual tweet in real time to observe useful information. As such, the goal is to influence different machine learning techniques for example, information classification and extraction to conduct the job repeatedly.

Related Issues

Demonstrating Effective Flexible Agent Coordination of Teams through Omnipresence (DEFACTO) facilitates the high standard AI, 3D visualization and human-interaction reasoning into a high-fidelity system for training responders. By providing the responders interaction with the coordinating agent team in a complex environment, the responder can get experience and draw valuable lessons that will be applicable in the real world.

Locating tactical and actionable information in real time within speedily growing information is challenging for several reasons. For example, performing information extraction on short bursts of text (e.g., on 140-character tweets) is more difficult than performing the similar task on large documents such as blog posts news articles. Furthermore, study has revealed that pre-trained classifiers drop in classification accuracy when used in different but similar disasters (Imran, Elbassuoni, Castillo, Diaz, & Meier, 2013). This requires learning and training new classifiers using fresh training data every time a crisis occurs.

During disasters, social media messages make available real-time awareness information that enables crisis responders to be more effective in relief efforts. However, different emergency response agencies are interested in different types of messages.

Role of machine intelligence: Conventional information processing cannot be used in this model, as disaster responders cannot wait to collect information and curate and categorize it online. As a substitute, responders and other stakeholders require real-time insight and intelligence as the disaster unfolds.

Role of human intelligence: When trying to perform non-trivial tasks, machines merely are not capable of high accuracy. Human interference is required to verify or correct the machine output. Use of human intelligence is the gap for the jobs that cannot be automated; for instance, providing input labels and validating the machine's output are among the types of human interventions.

Combined intelligence: investigating each individual message is challenging due to the scale of information posted on Twitter, which goes beyond the processing capability of humans. Nowadays, different technologies are available to disaster managers and initial responders. Drones, sensors and robots can provide accurate information in real time on damaged buildings and landscapes, making rescue efforts safer and faster. Augmented and virtual reality is used to increase training with more practical environments (Sebillo, Vitiello, Paolino, & Ginige, 2016, Julie, Elsa, & Murray, 2019). Serious games are used to raise awareness of roles and responsibilities of stakeholders participating in crisis management (Di Loreto, Mora, & Divitini, 2012). Computer simulation facilitates decision makers with analytical tools for evacuations that pragmatically model human behaviors.

Artificial Intelligence and Disaster Management: Nepal Scenario

AI integrates the data-crunching competence of high-speed computers with complex mathematical algorithms to rapidly analyze large volumes of information for valuable insights. Used to improve everything from medical diagnoses to product marketing campaigns, AI provides EM professionals awareness, such as complete damage assessment within a few minutes after disaster. The proprietary algorithms that strengthen seismic concern in the case of earthquake regard as the design and age of buildings, how they fared in previous earthquakes and surrounding soil conditions that influence risk levels. The platform analyzes other critical factors, such as population density to create heat maps showing the high vulnerable areas (Artificial Intelligence 2019; International Bank for Reconstruction and Development, 2018).

The information helps emergency management (EM) professionals and others to prioritize responses and utilize resources. After disaster, the algorithms continuously update and improve ongoing resource deployments as field reports, social media posts and communications from the EM ecosystem. As seismic concern is a critical tool during disaster, it can enhance training for future responses. Introducing pragmatic simulations based on geophysical data for table top exercises and full-scale drills, seismic concern keeps teams sharp and assists them identify gaps in response plans. The scenarios also provide EM professionals information they can take to local businesses, hospitals and schools to help recover emergency strategies and increase community safety (Artificial Intelligence, 2017).

Major role of AI in disaster management is highlighted as follows:

- Life and property of different people have been severely affected
- Several precautions and measures can be taken to give early alarm systems
- Concepts of Internet of Things (IoT) and AI would be very useful in such cases
- Timely Prediction of Earthquake, flooding and landslide
- Prediction of Radiation level at different geographical region of Nepal
- Timely precaution can be taken

Terms like artificial intelligence, machine learning, big data and deep learning are crucial for disaster management.

Machine Learning Terminologies

Feature: Characteristics used to describe the data samples and predict the output variable: Set of samples with known features and class labels. This set of labeled samples should be divided into training, validation and testing.

Training dataset: learn the relevant patterns in the features which are relevant to predict the output variable.

Validation dataset: Used to help determine the best model parameters.

Testing dataset: A set of labeled unseen data.

Class: splitting of a dataset into two or more groups of data that share some common characteristic.

Glacial lake output flood is another disaster phenomenon frequently happening in Nepal. The details of cause and losses are highlighted in Table 8.1. Also the distribution of glaciers in the river basins of Nepal is shown in Table 8.2.

Attributes for glaciers are classified as follows:

Table 8.1 Glacial lake outburst floods (GLOF) in Nepal (Ives, Shrestha, & Mool, 2010)

Entirely within Nepal					
	Date	*River basin*	*Lake*	*Course*	*Losses*
1 N	450 years ago	Seti Khola	Machhapuchchhre	Moraine collapse	Pokhara valley covered by 50-60 m deep debris)
2 N	3 Sep 77	Dudh Koshi	Nare	Moraine collapse	Human lives, bridges, others
3 N	23 Jun 80	Tamor	Nagma Pokhari	Moraine collapse	Villages destroyed 71 km from source
4 N	4 Aug 85	Dudh Koshi	Dig Tsho	Ice avalanche	Human lives, hydropower station, 14 bridges etc.
5 N	12 Jul 91	Tama Koshi	Chubung	Moraine collapse	Houses, farmland etc.
6 N	3 Sep 98	Dudh Koshi	Tam Pokhari	Ice avalanche	Human lives and more than NRs. 156 million
7 N	15 Aug 03	Modi River	Kabache Lake	Moraine collapse	Not known
8 N	8 Aug 04	Modi River	Barun Khola	Moraine collapse	Not known
9 N	Unknown	Arun	Barun Khola	Moraine collapse	Not known
10 N	Unknown	Arun	Barun Khola	Moraine collapse	Not known
11 N	Unknown	Dudh Koshi	Chokarma Cho	Moraine collapse	Not known
12 N	Unknown	Kali Gandaki	Unnamed (mustang)	Moraine collapse	Not known
13 N	Unknown	Kali Gandaki	Unnamed (mustang)	Moraine collapse	Not known
14 N	Unknown	Mugu Karnali	Unnamed (Mugu Karnali)	Moraine collapse	Not known

1. Unique identifier (basin and sub-basin name, glacier name, latitude, longitude, and highest, lowest and mean elevations
2. Physical parameters (area, length and orientation of glacier)

Attributes for hazard assessment are described as follows:

Table 8.2 Distribution of glaciers in the river basins of Nepal (Ives et al., 2010)

	2001 glacier inventory			*2010 glacier inventory*	
Basin	*No. of glaciers*	*Total area (sq. km)*	*Mean area (sq. km)*	*No. of glaciers*	*Total area (sq. km)*
Koshi	779	1410	1.81	843	1180
Gandaki	1025	2030	1.98	1337	1800
Karnali	1361	1741	1.27	1461	1120
Mahakali	87	143	1.65	167	112
Total	3252	5324	1.64	3808	4212

1. Large size and rapid expansion
2. Increase in water level
3. Intermittent activity of supra-glacial lakes
4. Position in relation to moraines and associated glacier
5. Dam condition
6. Glacier characteristics
7. Physical conditions of surroundings

A Case Study of Flood Damage Prediction

A flood event in the Netherlands in 1993 and six flood events in Germany between 2002 and 2013 were widely covered in media and they affected lives of people in great extent. Bayesian networks and regression random forests model were constructed and the result shows that the models which are trained using heterogeneous data (i.e., flood events with various characteristics) have a higher performance. A hybrid method has higher accuracy.

Koshi River in Nepal

- Combination of seven small rivers
- Catchment area 60,400 sq. km
- 150 km inside Nepal
- Flows from North Himalayas to Bihar, India in south
- Danger Level: 150,000 cusecs, in 2018, 215,970 cusecs.

Sensors can be placed in different parts of the road to detect flood. Wireless Sensor Networks can be placed in different parts of the road to detect flood. The sensors node comprises of Doppler's effect in the water flow measurement.

Data Collection

Rainfall data from Meteorological Forecasting Division, Department of Hydrology and Meteorology Nepal was collected (Source: http://www.mfd.gov.np), Nepal Disaster Risk Reduction Portal (Source: http://www.drrportal.gov.np/). Early warning of flooding provides real-time solution of disaster. The water flow and water level were measured during data collection. Following system was developed for forecasting the flood disaster (Fig. 8.3).

Data Visualization

Data is visualized in the form of scatter diagram or scatter diagram (Fig. 8.4). While developing the system the source code was written in Python so that data it would ease in data analytics. The proposed model of overall system is shown in Fig. 8.5.

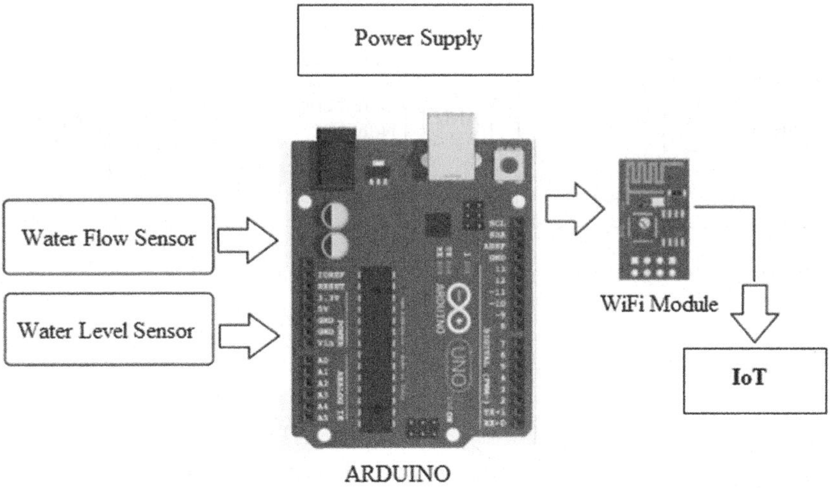

Fig. 8.3 Flood detection system (Early Flood Detection System, 2019)

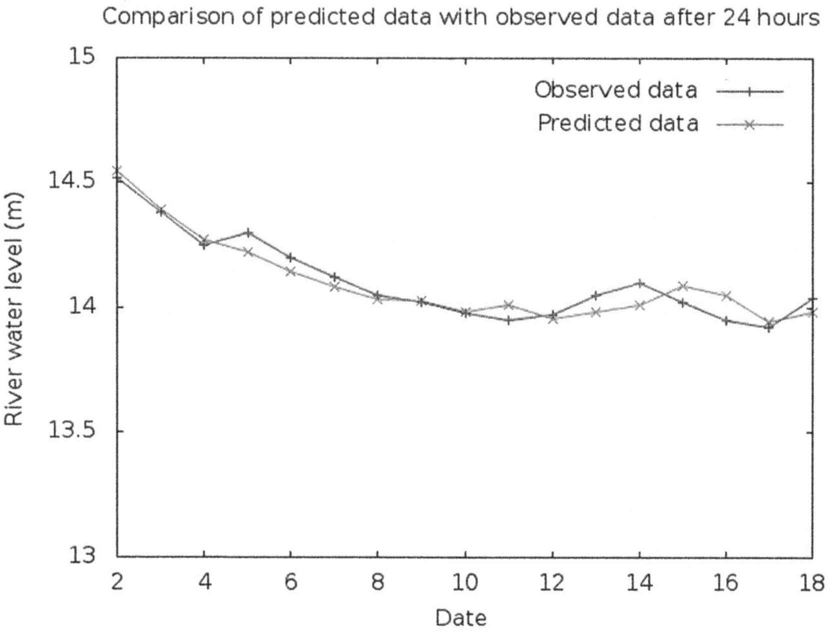

Fig. 8.4 Comparison of predicted data (Paul & Das, 2014)

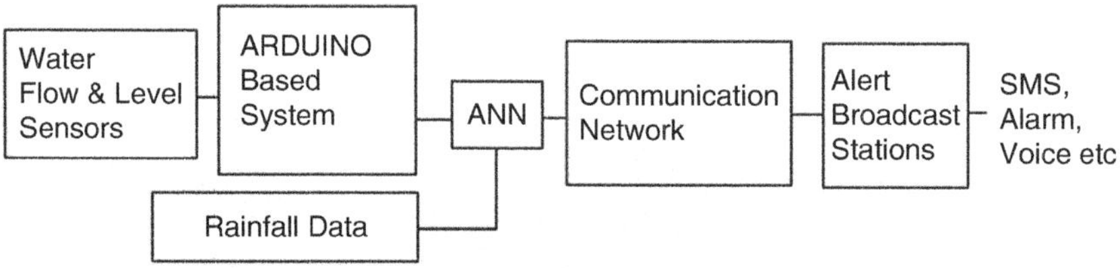

Fig. 8.5 Proposed model

The overall result is summarized as follows:

Radar pluvial flooding identification for drainage system (RAPIDS)	
Two artificial neural networks	
Input 1: Radar data	Input 2: Rainfall prediction
Output 1: Rainfall prediction	Output 2: Flood severity prediction

Flood Map Generation

When the discharge level reaches the critical level, warning is generated in the system and user generates a flood map following the warning message.

Water Level Prediction for Flash Flood

1. To use fuzzy inference system to predict the water level for flood prediction.
2. To send the warning message to the Disaster Management Center in case of flash flood for information dissemination.

Rapid Earthquake Assessment from Satellite Imagery

1. To determine the number of damaged buildings in the affected areas using satellite imagery.
2. To provide rapid initial assessment about the affected areas

Unmanned Aerial Vehicle (UAV) Based Emergency Management

- Recognizing sound patterns ("Help") related to disaster
- Locating person calling for help
- Implementation of Speech recognition algorithm and noise filtering

Recommendations from This Study

1. A more human centered approach is needed if crisis-warning apps are to be used effectively and with a good response from the population
2. AI has an important role to play in disaster management
3. Human behavior and in particular social cohesion plays a critical role in managing crisis situations
4. Evacuation should not only be studied in currently operational buildings, but software is for architects in the design phase.

Conclusion

AI techniques offer powerful tools for the development of disaster response and management systems. The technologies of robotics, ontology and semantic web, and multi-agent systems can be useful to solve the problems of crisis response. Social media platforms like Twitter receive an overwhelming amount of situational awareness information. For emergency response, real-time disaster insights are important. The success of any AI platform is directly tied to the proprietary data management and analytical capabilities engineered into the underlying algorithms. But effectiveness also hinges on other key technology components for enhancing EM team success.

References

Artificial Intelligence: A Game Changer for Emergency Response. The Power of Artificial Intelligence When It Matters the Most, Available online at: www.oneconcern.com/aboutus, Accessed 20 Mar 2019.

Di Loreto, I., Mora, S., & Divitini, M. (2012, June). Collaborative Serious Games for Crisis Management: An Overview. In *Enabling Technologies: Infrastructure for Collaborative Enterprises (WETICE), 2012 IEEE 21st International Workshop on* (pp. 352–357).

"Disaster Management", http://esw.w3.org/topic/DisasterManagement. Accessed 20 Mar 2019.

Early Flood Detection System. Available online at: https://nevonprojects.com/early-flood-detection/. Accessed 4 Mar 2019.

Geography of Nepal, 2020, Available online at https://en.wikipedia.org/wiki/Geography_of_Nepal. Accessed 4 Mar 2020.

David Greer, Phillip M. C. Kerrow, Jo Abrantes. (2002), "Robots in Urban Search and Rescue Operations", roceedingsACRA'2002, Auckland, November, pp. 25–30. McKerrow, P.J.

Imran, M., S. M. Elbassuoni, C. Castillo, F. Diaz, and P. Meier. 2013. Extracting information nuggets from disaster-related messages in social media. In *Proceedings of ISCRAM*, Baden-Baden, Germany.

International Bank for Reconstruction and Development/International Development Association, 2018.

Ives, J. D., Shrestha, R. B., & Mool, P. K. (2010). *Formation of Glacial Lakes in the Hindu Kush-Himalayas and GLOF Risk Assessment.* Kathmandu: ICIMOD.

Julie, D., Elsa, N., & Murray, T. (2019). ICT and Artificial Intelligence for Crisis and Emergency Management. Proceedings of the 52nd Hawaii International Conference on System Sciences, pp. 626–628.

Daniel Massaguer, Vidhya Balasubramanian, Sharad Mehrotra, and Nalini Venkatasubramanian. MultiAgent Simulation of Disaster Response. *AAMAS'06* May 8–12 2006, Japan.

Mehrotra, S., Butts, C., Kalashnikov, D. V., Venkatasubramanian, N., Rao, R., et al. (2004). Project Rescue: Challenges in Responding to the Unexpected. *SPIE, 5304*, 179–192.

Paul, A., & Das, P. (2014a). Flood Prediction Model using Artificial Neural Network. *International Journal of Computer Applications Technology and Research, 3*(7), 473–478.

Sebillo, M., Vitiello, G., Paolino, L., & Ginige, A. (2016). Training Emergency Responders Through Augmented Reality Mobile Interfaces. *Multimedia Tools and Applications, 75*(16), 9609–9622.

Gajendra Sharma received Ph.D. in information systems engineering from Harbin Institute of Technology, China. He did his postdoctorate from Dalian University of Technology, China. He is Associate Professor of the Department of Computer Science and Engineering at Kathmandu University, Nepal. He worked as Associate Professor at Liaoning University of Technology, China from 2011 to 2014. His interesting areas of research are crisis and emergency management, information systems and technology, and policy implementation. He received Nabil Science and Technology award from Nepal Academy of Science and Technology. He has published more than 90 papers in journals and international conference proceedings.

Subarna Shakya is Professor of Computer Engineering at Department of Electronics and Computer Engineering, Pulchowk Campus, Institute of Engineering, Tribhuvan University and also Visiting Professor in Brown University of Rhode Island, USA. He has served as head of department, as well as Executive director of National information technology center, Government of Nepal. He has also served as coordinator of Erasmus Mundus Program. He has delivered key note speech in national and international conference and workshop. He was awarded young scientist award from Nepal academy of science and technology as well as education leadership ward 2017 and award for outstanding contribution to education 2018 in word CSR day and world sustainability. He has published more than 100 research papers. His keen interest of research is in e-government system and security, computer systems simulation and modeling, cloud computing and energy efficiency, information system and software engineering, computer architecture and software engineering.

PART II

Government, Governance and Law

CHAPTER 9

Enhancing Accountability and Triadic Collaboration in Disaster Governance of Sri Lanka

Kokila Konasinghe

The Background

Sri Lanka has been facing different kinds of natural disasters including floods, droughts, and landslides commonly in the recent years, adversely affecting the lives of the people, their properties and livelihoods, exposing them to various kinds of dangers and resulting in losses in the national economy. The latest of such disasters is the Fall Armyworm (FAW) menace which according to the Minister of Agriculture, destroyed an estimated 20% of the agricultural cultivations in Sri Lanka by January 2019 (Parliamentary Debates, 2019). While these disasters appear to be natural in the outset, a careful analysis attributes them to adverse human activities including anthropogenic climate change, unsustainable land use and mismanagement of harm prevention strategies and failure to carry out full

K. Konasinghe (✉)
Department of Public and International Law, Faculty of Law, University of Colombo, Colombo, Sri Lanka
e-mail: kokila@law.cmb.ac.lk

T. V. V. Kumar, K. Sud (eds.), *AI and Robotics in Disaster Studies*, Disaster Research and Management Series on the Global South,
https://doi.org/10.1007/978-981-15-4291-6_9

environmental impact assessments for development projects among other things (Konasinghe, 2018). The role of the human activities in causing, triggering or exaggerating natural catastrophes places them outside the scope of non-preventable disasters and imposes a mandatory obligation on the government authorities and common citizenry not only to address the consequences of natural disasters but to take measures to predict and prevent such disasters.

The use of Artificial Intelligence (AI) in predicting disasters and responding to disaster situations has proved to be effective than conventional disaster management methods (Imran, Alam, Ofli, & Aupetit, 2017). The focus of this chapter is to analyse how the triadic collaboration of human rights, precautionary principle and sustainable development encourages and imposes an obligation on the decision-making authorities in Sri Lanka to utilise AI in disaster risk reduction and management.

The Role of Artificial Intelligence in Disaster Risk Reduction and Management

AI as defined by one of its founding fathers, "is the science and engineering of making intelligent machines, especially intelligent computer programs" (McCarthy, 2007). Advancement of AI has positively affected disaster risk reduction (DRR) and disaster management (DM) efforts in numerous ways.

Disasters create uncertain situations which require an immediate reaction and assistance. However, the determination of the gravity of the disaster, urgent needs and location of the affected population, ability to access shelters and recognition of obstructed roads have proved to be a dauting task in the absence of accurate, trustworthy information even at situations where pre-determined DM plans are in play. AI can be used to process real-time information in disaster situations (Imran et al., 2017). Image recognition and classification capability of AI can be used to filter and identify satellite images of injured people, damaged infrastructure including roads, buildings and bridges which can be used to ascertain the severity of the damage and the best way to access the victims (Lahoti, 2018). AI can also combine data from multiple resources, remove unreliable data and identify informative sources to generate heat maps which shows the areas in need of urgent assistance (Lahoti, 2018). The efficiency and accuracy of the information provided in this process can play a vital role in managing and responding to disasters and saving lives and properties of the people.

Secondly, telephone lines of emergency relief services (for example 911 in USA and 119 in Sri Lanka) can get extremely overloaded in a disaster situation. It deprives the ability of the emergency service providers to receive and respond to all requests for assistance. The use of AI in these services enables the receipt, proper categorisation and prioritisation of the requests for help and appropriate interaction with the callers while ingesting data not only from voice calls but also from text messages, videos and pictures (Lahoti, 2018).

Thirdly, AI can be used to analyse the data reported in social media websites such as Facebook, Twitter and Instagram on disasters either by victims themselves or by people acting as citizen journalists. Existing research studies reveal that people tend to report useful information relating to disasters in social media including the locations of victims, urgent needs, areas to which the prioritised attention shall be given, communities who have not received adequate assistance etc. (Mauroner & Heudorfer, 2016). This role of social media was well-apparent in the flood disasters taken place in Sri Lanka in 2016 and 2017 (Ahamed, 2016), (Vanessa, 2017). AI can collect and analyse these real-time data, filter genuine information and distribute among the relief workers to enable them to access the right place at the right time with the right aid without wasting time which is a crucial determinant in disaster situations.

AI can be used not only in disaster reciliation but also in the prediction and forecasting of probable and possible disasters which can be the difference between life and death for thousands of people (Joshi, 2019). AI systems can be developed with the help of different datasets including seismic data, rainfall records and ash particles to predict earthquakes, tsunami, floods, volcano eruptions and other natural disasters (Joshi, 2019) and to issue automated warnings to the possible victims and relevant authorities. Through the accurate prediction of the disasters thousands of lives can be saved while minimising economic and property losses.

It is clear that AI provides a better chance of saving lives and properties either through proper channelling of aid during disasters or through the prior notification of impending disasters. AI has been used in actual practice in disasters like hurricane Maria in 2017 and hurricane Florence in 2018 with promising results (Mauroner & Heudorfer, 2016), (Clark, 2018). While the world is striving in the path of enhancing the efficacy of DRR and DM mechanisms through the use of AI, it is important to analyse what obligation does this positive development impose on the decision-making bodies in Sri Lanka.

The Legislative Framework on Disaster Risk Reduction and Management in Sri Lanka

The main legislation in force in Sri Lanka with regard to disaster management is Sri Lanka Disaster Management Act, No. 13 of 2005 which provides for the establishment of the National Council for Disaster Management, the Disaster Management Centre, appointment of technical advisory committees, the preparation of disaster management plans, the declaration of a state of disaster and award of compensation. The Act was subsequently amended in order to fill the existing gaps and introduce pragmatic approaches ensuring disaster preparedness, early warning generation and dissemination, mitigation measures, early warning response, providing relief services and arranging recovery (Sri Lanka Disaster Management (Amendment) Act, No. 13 of 2005).

The umbrella legislation on environmental law in Sri Lanka, the National Environmental Act, No. 47 of 1980 as amended also provides for the management and mitigation of disasters through environmental protection and conservation, maintenance of environmental quality and achievement of sustainability. The Act takes a holistic approach to prevent environmental pollution and prohibits water pollution through sections 23 (G), (H), and (V), atmospheric pollution through sections 23 (J) and (K), soil pollution through sections 23 (M) and (N) and noise pollution through sections 23 (P), (Q) and (R). Part IV A of the Act deals with "environmental protection" and states that discharge, deposition or emission of pollution causing waste into the environment shall be carried out only under the authority of an environmental protection licence and in accordance with the standards and criteria prescribed in the Act. Part IV C of the Act mandates an environmental impact assessment to be carried out for development projects which are either large scale or located in environmental sensitive areas. These provisions ensure that the environment is preserved from adverse implications of development process and human activities and therefore natural disasters will not be generated or aggravated due to environmental change, degradation or vulnerability.

However, none of these legislations specifically encourages the use of AI as a means of immediate disaster response and management. On the other hand, both of these legislations do not take a human rights approach towards DRR and DM and therefore do not highlight the needs of the human beings affected and assist in creating a holistic framework for necessary action (Konasinghe, 2018). While there have been research works carried out in Sri Lanka which seek to determine the potential of AI in effective DRR and DM, such attempts are not encouraged or put into actual implementation through the legislative framework in Sri Lanka.

The Role of the Government of Sri Lanka in Disaster Situations

The mandatory and obligatory role of the government in disaster mitigation and management has not been adequately elaborated in the Sri Lankan context. Therefore, it is crucial to establish that the role of the government in preventing, managing and extending humanitarian support in disaster situations is not a mere charity or a welfare activity but a Constitutional mandate.

According to article 27 (1) of the 1978 Constitution of Sri Lanka, Directive Principles of State Policy shall guide the parliament, the president and the cabinet of ministers in the enactment of laws and the governance of Sri Lanka for the establishment of a just and free society. Article 27 (2) states that the State is pledged to establish in Sri Lanka, a Democratic Socialist Society, the objectives of which include the full realisation of the fundamental rights and freedoms of all persons and the realisation by all citizens of an adequate standard of living for themselves and their families, including adequate food, clothing, housing and continuous improvement of living conditions. Disasters, whether natural or man-made do not make an exception to this obligation imposed upon the State by the Constitution and more rationally it shall carry a greater weight during disaster situations. Moreover, sub-article 14 states that the State shall protect, preserve and improve the environment for the benefit of the community. This implies the obligation of the State to protect people from preventable natural disasters.

In terms of article 29, none of these provisions are enforceable before any court or tribunal. It has nevertheless been established by the judiciary in Sri Lanka that the Directive Principles of State Policy do not become entirely redundant by virtue of article 29. As held in *Environmental Foundation Ltd v Mahaweli Authority of Sri Lanka* (2010), although it is expressly declared in the Constitution that the Directive principles do not confer or impose legal rights or obligations and are not enforceable in any Court or Tribunal, the Directive principles are linked with the public trust doctrine and should guide state functionaries in the exercise of their powers (p. 19). As held in *Ravindra Gunawardena Kariyawasam v Central Environmental Authority and Others* (2019), "the Directive Principles of State Policy are not wasted ink in the pages of the Constitution. They are a living set of guidelines which the State and its agencies should give effect to" (p. 50). In terms of these decisions, the role of the government in Sri Lanka in preserving the environment to prevent natural disasters, providing the victims with humanitarian aid and protecting the fundamental rights and freedoms of the people during disasters goes way beyond an act of sympathy or charity.

According to article 3 and 4 of the Constitution, the organs of the government are exercising the powers of people, therefore any exercise of such powers shall take into account the ultimate benefit of the public. As held by Eva Wanasundara PC J, "discretions are conferred on public functionaries in trust for the public, to be used for the good of the public" (*Premalal Perera v Tissa Karaliyadda*, 2016, p. 5). As held in the *Water's Edge* case (2007) "all facets of the country …are to be handled and administered under the stringent limitations of a trusteeship posed by the public trust doctrine and must be used for the benefit of the entirety of the citizenry of the country" (p. 341). Thus, if the use of AI appears to be the most effective and efficient way to predict and prevent disasters prior to their occurrence and to accurately locate and channel aid to the victims during disasters, the government is under an obligation to make use of AI appropriately. This obligation is further heightened when considered in line with the triadic collaboration of human rights, sustainable development and precautionary principle.

The Human Right to Be Protected and Relieved from the Disasters

As pointed out by George Kent, international human rights law does not explicitly address the right to protection and relief from disasters, but this objective is clearly implied (Kent, 2013). Three major articles in the Universal Declaration of Human Rights adequately imply the right to protection from disaster losses. Article 3 states that everyone has the right to life, liberty and security of person, article 17 states that everyone has the right to own property alone as well as in association with others and article 25 (1) states that everyone has the right to a standard of living adequate for the health and well-being of himself and of his family, including food, clothing, housing and medical care and necessary social services, and the right to security in the event of unemployment, sickness, disability, widowhood, old age or other lack of livelihood in circumstances beyond his control (Rawinj, 2013). Human beings as right-holders can claim their rights and demand from duty-bearers for their fulfilment. The State as the main duty-bearer has both the moral and legal obligation to respect, protect, facilitate, and fulfil human rights. The State has the obligation to prevent loss of lives, including losses of economic and social assets, and to prevent other human rights violations, whether caused by human or natural forces (Rawinj, 2013). The UN International Law Commission in its draft articles on the Protection of Persons in the Event of Disasters adopted in 2014 made a clear reference to the link between human rights and DM. According to article 6 of the draft articles, persons affected by disasters are entitled to respect for their human rights.

The Constitution of Sri Lanka clearly lags behind international standards when considering the rights that can be invoked to establish a human right to protection and relief from the disasters. However, the fundamental rights regime in Sri Lanka is not entirely devoid such rights. In the case *Sriyani Silva v Iddamalgoda, OIC, Police Station Payagala and Others* (2003), the Supreme Court of Sri Lanka created an implied right to life holding that the people shall have the right to life at least in the sense of mere existence. The Constitution also recognises the rights of the people to movement and to choose residence. These rights are not suspended in disasters. The government may have limited resources, capabilities and capacities during disaster situations to facilitate the full realisation of these rights, but they cannot entirely evade such responsibility and deny the rights of the people. Nevertheless, no matter how meagre their resources may be, all governments have an obligation to take positive action to protect lives, and to assure an adequate standard of living, not only in normal times but also in conditions of acute crisis (Kent, 2013).

On the other hand, impliedly evolved fundamental right to environment in a wave of cases including *Eppawela phosphate mining case* (2000), *Galle Face Green case* (2004), *Watte Gedara Wijebanda case* (2007), *noise pollution case* (2005), *Waters' Edge Case* (2008) and the most recent *Ravindra Gunawardena Kariyawasam v Central Environmental Authority and Others case* (2019) confers upon the people a right to have the environment preserved as a precautionary measure against natural disasters.

Human Rights are maximum claims on society [which elevate the necessity of safeguarding the environment and common citizenry against disasters] above a mere policy choice that may be modified or discarded at will (Shelton, 2010). Therefore, the State of Sri Lanka has an obligation to make use of the AI in DRR and DM in ensuring the highest safeguards to the human rights of the people.

Sustainable Development and Prevention of Disasters

Sustainable development can be defined as the development that meets the needs of present without compromising the ability of future generations to meet their own needs (Brundtland, 1987). According to principle 04 of the Rio Declaration in 1992, in order to achieve sustainable development, environmental protection shall constitute an integral part of the development process and cannot be considered in isolation from it. Principle 25 of the same declaration lays down that, peace, development and environmental protection are interdependent and indivisible. International judicial pronouncements in *Gabcíkovo-Nagymaros project case* (1997), *shrimp-turtle case* (1998) and *pulp mills case* (2010) clearly indicate that the principle of sustainable development has been well recognised in the international legal regime at least as an objective. Accordingly, all the states shall respect their obligation to carry out development activities in an environmentally sound manner.

The principle of sustainable development was recognised for the first time by the judiciary in Sri Lanka in the *Eppawela phosphate mining case* (2000) which specifically referred to the sustainable development principle embodied in the Stockholm Declaration and the Rio Declaration. In the case Amerasinghe J. laid down in clear and unambiguous terms that the sustainable development principle recognised in Stockholm and Rio Declarations "will be binding if they have either expressly enacted or become a part of the domestic law by adoption by the superior court of record and Supreme Court in particular in its decisions" (p. 274). The court further held that the human development paradigm needs to be placed within the context of our finite environment so as to ensure the future sustainability of the mineral resources and of the water and soil conservation ecosystems of the Eppawela region and of the North Central Province and Sri Lanka in general (p. 279). In the subsequent *Watte Gedera Wijebanda case* (2007), Thilakawardane J, following *Eppawela phosphate mining case* (2000) held that the Stockholm Declaration and Rio Declaration though not legally binding upon government, constitute a form of soft law, the importance and relevance of which must be recognised when reviewing executive action vis-à-vis the environment. The Court in this case adverted to Principle 1 of the Rio Declaration that human beings are the centre of concern for sustainable development and they are entitled to a healthy and productive life in harmony with nature (p. 21).

The principle of sustainable development compels the authorities to make development decisions with due regard to their long-lasting consequences and thus seeks to prevent natural disasters stemming from or triggered by unsustainable decision-making and improper use of natural resources. The use of AI provides better means to achieve sustainable development goals including good health and well-being, gender equality, clean water sanitation, reduction of inequality and sustainable cities and communities, most particularly during disasters or through proper prevention of disasters. On one hand, AI upholds sustainable development goals through facilitation of proper channelling of assistance; food, water and healthcare facilities to the victims of disasters without discrimination or inequality. On the other hand, AI can anticipate and prevent disasters which ensures that a set of people will not bear a disproportionate burden of anthropogenic natural disasters and their access to basic human needs will not be compromised as a result. Moreover, disasters can hinder the achievement of long-term sustainable development goals through diversion of attention and resources of the decision-making authorities to short-term goals associated with disasters including providing essential public health and safety services, restoring interrupted utility and other essential services, re-establishing transportation routes, and providing food and shelter for those displaced by the incident (Esnard & Sapat, 2014). Prevention of disasters can maintain an uninterrupted focus on the achievement of sustainable development goals and keep sustainable development plans on operation.

Precautionary Principle in Averting Disasters in Advance

Precautionary principle upholds the view that prevention is better than cure, i.e. it is better to prevent environmental damage in the first place than allow it to occur and then trying to repair the damage. It does not mean that damage to the environment may not be sanctioned. Rather, it means that decision-makers ought, at the very least, to make themselves aware of the potential effects of what they are sanctioning, in order to be able to determine what level of environmental change or risk of change is necessary (McIntyre & Mosedale, 1997). To take precautionary actions the possible and probable environmental impacts of a certain activity shall be ascertained in light of existing scientific knowledge. Rio Declaration has broadly recognised the precautionary principle and laid down that in order to protect the environment, the precautionary approach shall be widely applied by States according to their capabilities. According to principle 15 of the Rio Declaration, where there are threats of serious or irreversible damage, lack of full scientific certainty shall not be used as a reason for postponing cost-effective measures to prevent environmental degradation. The inclusion of the principle in the Rio Declaration marked the elevation of the principle to the status of a core principle of international environmental law making (Freestone, 1994). Burnett-Hall observes that the prevention principle has a meaningful role where foreseeable environmental harm is likely as a consequence of proposed action; the precautionary principle comes into play where not only the likelihood of harm, but also its nature and extent, may all be uncertain (Burnett-Hall & Jones, 2012). Precautionary principle plays a definitive and undeniable role in sustainable development.

Precautionary principle has been relied upon by the international courts in a number of judicial decisions. In *Gabcíkovo-Nagymaros project case* (1997), Hungary correctly perceived precautionary principle as a link between principle of corporation and state responsibility. In *French nuclear tests case* (1995), New Zealand contended that France is under an obligation to act in accordance with the precautionary principle. While the majority of the court dismissed New Zealand's request, the dissenting judgements of judges Palmer, Koroma and Weeramantry emphasised the significance of the precautionary principle. In the *Nirex case*, Ireland raised the precautionary principle arguing that it imposes duties on the British government in relation to the application (Marr, 2015). In *Southern bluefin tuna case* (2000), the International Tribunal for the Law of the Sea considered the precautionary principle without expressly relying on it and incorporated this principle in its order.

The precautionary principle was cited by the Supreme Court of Sri Lanka, in the *Eppawela phosphate mining case* (2000). In the case Amerasinghe J upheld the more progressive approach to the precautionary principle as elaborated in the Rio declaration and laid down in clear terms that "if ever pollution is discerned, uncertainty as to whether the assimilative capacity has been reached should not prevent measures being insisted upon to reduce such pollution from reaching the environment" (p. 287).

The precautionary principle emphasises the necessity of foreseeing natural disasters and taking measures to prevent them. It further emphasises the need of not postponing cost-effective measures to protect the environment. In this regard the role that can be played by AI is extremely critical. As mentioned earlier in this chapter, base data can be fed into the AI systems and can predict possible disasters. This mechanism can save thousands of lives and properties that could otherwise have lost due to disasters and prevent irreversible damage to the environment.

Conclusion

The role of AI in DRR and DM is crucial and proved to bring positive results. While it is often perceived in the Sri Lankan context that the government authorities do not hold a mandatory obligation to prevent and address consequences of natural disasters because they are usually regarded as occurrences beyond control, the triadic collaboration of human rights, sustainable development and precautionary principle suggests that this duty goes beyond a mere act of sympathy or charity. The rights of the people to live and environment impliedly suggest that they have a right to be protected and relived from natural disasters. The governments on the other hand are under a mandatory obligation not only to channel assistance to the victims of the disasters but also to foresee possible disasters and to take precautionary measures. The rights of the people do not become suspended during disasters and the relevant authorities carry an obligation to safeguard the rights of the people irrespective of the situation, during normal times and disasters. The authorities shall carry out this duty vested within them in the best way available, and the use of AI to predict and prevent disasters and to react in the disaster situations provides one of the best ways to achieve this objective.

References

Ahamed, H. (2016). *Hats Off to Our WhatsApp Warriors.* Retrieved from https://groundviews.org/2016/05/22/hats-off-to-our-whatsapp-warriors/

Al Haj MTM Ashik v RPS. Bandula, OIC Weligama (the noise pollution case) SC (FR) Application No. 38/2005 (2007).

Brundtland, G. (1987). *Our Common Future: The World Commission on Environment and Development.* Oxford: Oxford University Press.

Bulankulama v Secretary, Ministry of Industrial Development (Eppawela phosphate mining case) 3 SLR 242 (2000).

Burnett-Hall, R., & Jones, B. (2012). *Burnett-Hall on Environmental Law.* London: Sweet & Maxwell.

Case Concerning Pulp Mills on the River Uruguay (Argentina v. Uruguay) (pulp mills case) ICJ 60 (2010).

Clark, C. T. (2018). *Visualizing Hurricane Florence.* Retrieved from https://www.forbes.com/sites/charlestowersclark/2018/09/18/visualizing-hurricane-florence/#462d6712f948

Environmental Foundation Limited v Urban Development Authority (Galle Face Green Case) SC (FR) No. 47/2004.

Environmental Foundation Ltd v Mahaweli Authority of Sri Lanka 1 SLR 1 (2010).

Esnard, A. M., & Sapat, A. (2014). *Displaced by Disaster: Recovery and Resilience in a Globalizing World.* New York: Routledge.

Freestone, D. (1994). The Road to Rio: International Environmental Law after the Earth Summit. *Journal of Environmental Law, 6, 193*(2).

French nuclear tests case (New Zealand v France) ICJ Reports 28 (1995).

Gabcíkovo-Nagymaros Project Case (Hungary v Slovakia), ICJ Reports 78 (1997).

Imran, M., Alam, F., Ofli, F., & Aupetit, M. (2017). *Enabling Rapid Disaster Response Using Artificial Intelligence and Social Media.* International Roundtable on the Impacts of Extreme Natural Events: Science and Technology for Mitigation (IRENE) workshop. Colombo, Sri Lanka.

Joshi, N. (2019). *How AI Can and Will Predict Disasters.* Retrieved from https://www.forbes.com/sites/cognitiveworld/2019/03/15/how-ai-can-and-will-predict-disasters/#4b76c3c55be2

Kent, G. (2013). The Human Right to Disaster Mitigation and Relief. *Environmental Hazard, 3*(3), 137.

Konasinghe, K. (2018, November). *The Need for a combined Sustainable Development and Human Rights Approach to Disaster Prevention and Management.* Paper presented at the annual research symposium of University of Colombo, Colombo. Retrieved from https://cmb.ac.lk/wp-content/uploads/ars_proceedings_2018.pdf

Lahoti, S. (2018). *AI to the Rescue: 5 Ways Machine Learning Can Assist During Emergency Situations.* Retrieved from https://hub.packtpub.com/ai-rescue-5-ways-machine-learning-can

Marr, S. (2015). *The Precautionary Principle in the Law of the Sea.* Leiden: Martinus Nijhoff.

Mauroner, O., & Heudorfer, A. (2016). Social Media in Disaster Management: How Social Media Impact the Work of Volunteer Groups and Aid Organisations in Disaster Preparation and Response. *International Journal of Emergency Management, 12*(2), 196.

McCarthy, J. (2007). *What Is Artificial Intelligence.* Retrieved from http://www-formal.stanford.edu/jmc/whatisai.pdf

McIntyre, O., & Mosedale, T. (1997). The Precautionary Principle as a Norm of Customary International Law. *Journal of Environmental Law, 9*(2), 221.

Parliamentary Debates, 23.01.2019 Vol. 267 No. 6 Col. 587–592.

Premalal Perera and others v Tissa Karalliyadda and others SC FR Application No. 891/2009 (2016).

Ravindra Gunawardena Kariyawasam v Central Environmental Authority and Others SC FR Application No. 141/2015 (2019).

Rawinj, F. (2013). *Claiming the Human Right to Protection from Disasters: The Case for Human Rights-based Disaster Risk Reduction.* Retrieved from https://www.preventionweb.net/files/submissions/31225_righttodisasterprotection.pdf

Shelton, D. (2010). Human Rights and the Environment: Substantive Rights. In D. M. M. Fitzmaurice (Ed.), *Research Handbook on International Environmental Law* (p. 266). Cheltenham, UK: Edward Elgar Publishing Limited.

Southern Bluefin Tuna case (Australia and New Zealand v Japan) 39 ILM 1359 (2000).

Sri Lanka Disaster Management (Amendment) Act, No. 13 of 2005.

Sriyani Silva v Iddamalgoda, OIC, Police Station Payagala and Others SC FR No. 471/2000 59(2003).

Sugathapala Mendis v Chandrika Bandaranaike Kumaratunga (the Water's Edge case) SC FR 352/2007 60 (2008).

United States Import Prohibition of certain Shrimp and Shrimp Products (shrimp turtle case) WT/DS58/AV/R (1998).

Vanessa. (2017). *Flood Relief Sri Lanka 2017.* Retrieved from http://socialmedia.lk/flood-relief-sri-lanka-2017/

Watte Gedara Wijebanda v Conservator General of Forest and eight others SC Application No. 118/2004 (2007).

Dr. Kokila Konasinghe is a Senior Lecturer at the Department of Public and International Law in the Faculty of Law, University of Colombo, Sri Lanka, where she currently teaches Human Rights Law and Environmental Law. She is the Director of the International Office of the University of Colombo, the Course Director of the LL.M. in Environmental Law and Policy and the Co-ordinator of the Certificate Course in Environmental Law at the Faculty of Law of the University of Colombo. She was a research fellow at O.P. Jindal Global University, India and a teaching fellow at Erasmus University, Rotterdam in the Netherlands. Her research interests are in the North-South dimension in global environmental governance, cultural relativism and universality of human rights, disaster management related studies, and law and literature. She is an attorney at law of the Supreme Court of Sri Lanka and a Staff and Educational Association of UK (SEDA) accredited university lecturer. She received her LL.B. (Hons.) from the University of Colombo and LL.M. from Erasmus University, Rotterdam and her PhD from Keele University, UK.

CHAPTER 10

Artificial Intelligence and the Legal Response Paradigms in Disaster Management

Stellina Jolly and G. S. Moses Raj

Introduction: Disaster Management Challenges and Scope of Employing Artificial Intelligence

Disasters, whether it is natural or human-made, has always been part of human history. However, in the last two decades, the studies and reports indicate exponentially increased disasters in magnitude and intensity. In 2017 itself, disasters have cost humanity a whopping $306 billion worldwide (Shephered, 2018). The situation is predicted to assume dangerous proportions when we take into account the climate change phenomenon and the scientific assessment of the Intergovernmental Panel on Climate Change [hereinafter IPCC] reports which highlight the complex interlinkage between climate change and disasters (Verchick, 2016; IPCC, 2012).

The question in this context is to see whether the big data supported by AI can be instrumental in helping the society cope with disasters when

S. Jolly (✉) • G. S. M. Raj
South Asian University, New Delhi, India
e-mail: stellinajolly@sau.ac.in

T. V. V. Kumar, K. Sud (eds.), *AI and Robotics in Disaster Studies*,
Disaster Research and Management Series on the Global South,
https://doi.org/10.1007/978-981-15-4291-6_10

governments and agencies are struggling to coordinate effective disaster relief programs which are often random, fragmented and disorganized. In spite of its lack of familiarity, AI is a technology that is transforming every walk of life, and the employment of AI in disaster management offers tremendous potential for the society in implementing the Sendai Framework for Disaster Risk Reduction 2015–2030 and pursuing risk-informed sustainable development (ITU, 2019).

Firstly, the use of technologies like sensor data, satellite imagery, wireless frequency technology etc. increases the capacity of agencies to predict and assist the multi-dimensional understanding and mapping of risk in case of disasters (Meier, 2015). The AI could assist not only in general prediction but to pinpoint the most potential area and communities to be affected. This will be a real boost for informed decision making and planning. For instance, if it is known that sand bank of Yamuna river in Delhi could be hard hit, better planning can be articulated in terms of building norms, construction and settlement activities. Further, the utility of AI immediately in the aftermath of disasters can be crucial in reducing the mortality and losses (Series, 2018). AI technologies can underpin response management through drones, satellite imagery, chatbots, social media, etc. These technologies aided by adequate data can provide accurate information about damaged buildings, roads, flooded area and landscapes thereby prioritize responses and avoid wasted effort (Harris, 2018). Drones were deployed in the cyclone hit South Pacific archipelago nation of Vanuatu to assess the damage and devastation (ITU, 2019).

Further, when challenges emerge to combat disaster preparedness, the agencies involved in mitigating damage suffered can employ the integration of social media, crowdsourcing and collaborative tools into disaster relief systems (Gao, Barbier, & Goolsby, 2011). Coordination, accuracy, and security of interagency activities as a shared mission to effectively locate people and resources using geo-tagging to report actual spatial features becomes crucial to address distress. Can Artificial Intelligence (AI) leverage crowdsourcing as a tool to accelerate the relief process that augments a real-time crisis map to identify and respond to unexpected cases? Given the scale and complexity of a massive disaster, agencies receiving unstructured data from several sources will need computational sophistication to be analysed and quickly deploy support systems to evacuate, rehabilitate and restore conditions through state agents such as disaster relief management forces or the Army. As volunteers and victims use communication channels to report incidents and requests either through social networking websites or human connections, a robust database that has enough interoperability can help first responders, governmental agencies and NGOs become more proactive in deploying aid and rescue capabilities (Simon, Goldberg, & Adinia, 2015). For instance, in the Nepal earthquake in 2015, AI was used to analyse the tagged tweets to identify urgent needs, infrastructure damages, and resource deployment needs (Guay, 2015).

Thus, AI is not just a future vision but a present reality with mammoth potential for augmenting human capabilities in many ways. At the same time, like with the development of any new technology, there are paramount ethical and legal issues raised. In this context, this chapter will explore the ethical and legal concerns raised by AI in the application of disaster management. The focus will be on issues of personality, liability, responsibility, discrimination, and access. The chapter will assess the existing frameworks and will argue for an AI system based on human dignity at its core.

Ethical and Legal Concerns in the Application of Artificial Intelligence

The primary rationale and legitimacy behind the application of AI in disaster management are grounded in the assumption that their employment will make a difference for those who are affected or will be affected with disaster. At the outset itself, it is essential to mention that the ethical and legal issues arising out of AI cannot be put into watertight compartments and any overlapping between them is anything but inevitable. The discourse on the ethical and legal issues arising out of the applications of AI are in the initial stages and is still evolving.

The development of AI is based on big data and the first ethical challenge is to design and develop an AI with an ethical face projecting concern for human dignity while addressing core issues such as data privacy, discrimination, human and social welfare and the sustainability of the use of technology (West, 2018; Corea, 2018). Secondly, as there is every possibility for receiving data and sending data will be compromised by unruly elements and vested interest groups, stability and security of the system calls for a great deal of micro-level supervision and resistance to withstand attacks. Therefore, designing an AI system ethically is not enough—it must also resist unethical human interventions. Furthermore, AI systems have an additional vulnerability: inputs can be manipulated in small ways that can completely change decisions (Shaw, 2019). Humanitarian assistance can become automated leading to self-decision-making powers thereby creating new decision trees and unsupervised algorithmic innovations which will cause issues of data privacy as instructions gets sent out to workers through the medium of cyberspace (Schmitt, 2013). Enhancing solidarity of human lives through institutional protection in an era of exploding technological innovation poses deeper questions that need to be answered within the structure of human values and existential threats that are perceived and imagined when control is given away to a machine (Bostrom, 2014).

As AI and its welfare properties are rooted in computational sophistication and is difficult to decode once the task request is made, AI matures and proliferates autonomously and creates problems of democratising data through a decentralized governance system that encapsulates localization and ownership (Singh, 2018). Government agencies then have to be more vigilant about civil and criminal consequences of deploying AI for automating mass civilian exercises such as evacuation and resettlement policies. Then, to what extent can a legal theory posit human rights within the AI framework while public policy engagements using big data could be termed as a Faustian bargain of a social good in a digital society needs explanation (HKS, 2018). This becomes critically relevant in case of the possibility of discrimination percolating at the implementation of AI. AI relies on machine learning and the parameters incorporated. If the specific damage scenario has not been coded into the machine, it can delay the relief operations and even lead to discrimination. For instance, an AI system which can identify damage to wooden/cement buildings, with specific shapes and colours if employed to a new area may not identify buildings made of other materials, shapes or colours leading to discrimination and lack of effectiveness. The effectiveness of AI is identified in percentages as showing an accuracy of 90 or 99 percentage. The 'percentage' cannot be the representative of efficiency or optimality; what is to be seen is the inclusiveness and comprehensiveness which protects the human right concerns. It needs to be emphasised that disaster management is heavily influenced by international human rights law which imposes a positive obligation on the states to protect human rights. If the application of AI brings human rights concerns in the form of discrimination and privacy, AI can also play a relevant role in protecting the host of rights infringed during disasters including right to life, property etc. (Sommario & Venier, 2018).

As we become further embedded into an algorithmic society (Aneesh, 2009) through Internet of Things, one must be vary of surrendering information control to a system that is sentient as well as intellectually and morally superior to humans and possesses the power to converge into issues such as agency and autonomy, liability and punishment, privacy and surveillance and personal sovereignty as it applies to consumer robots, industrial automation, smart buildings, unidentified aerial vehicles and drones, and autonomous vehicles (Chopra & White, 2011). This intersection between law and technology can frame a dialogue to address systemic inequities of AI without dismissing it off as science fiction. It is here that Science and Technology Studies as a discipline makes one question on responsive innovation through a collective stewardship of science and humanity while looking critically at contingency and societal challenges to think of ways to democratise thinking about the future (Stilgoea, Owen, & Macnaghten, 2013).

AI technology is founded on access to big data for predictive analysis and insights. This will require access to data from multiple sources and jurisdictions and simultaneously poses a real threat to data privacy and raises legal questions of jurisdiction, choice of law and enforcement issues (Pagallo, Corrales, Fenwick, & Forgo, 2018). The issue of privacy has already been the focus of jurisdictions with Europe, China and India leading the way. There seems to be the emergence of a shared view when it comes to data ownership and control of data (Garnacho, Ferrers, & Solanas, 2009). The fundamental principles of privacy consist of getting clear protection and security notices. An individual should be informed about the purpose, use, sharing and retention of the data and granting her the right to informational self-determination (Blodgett-Ford, 2018). Further, the collected data should be retained only as long as necessary. In this context, the question of emphasis is on transparency, explainability and information sharing on how the AI is being used and works. People have a right to know the purpose for which the data is being collected and used. Along with the legal and regulatory framework, this will also require creating public awareness about the systems and data. In the Indian context, the task force set up to assess the challenges and potentialities of AI has highlighted the ethical and legal concerns of fairness, transparency and privacy concerns. With regard to privacy, the task force has recommended the establishment of data protection framework (GOI, 2018). The discussion paper developed by the NITI Aayog recommended the harmonization of Indian privacy laws to match the international standards (Aayog, 2018). The discussion paper also referred to the European Union's less invasive General Data Protection Regulation (GDPR) guidelines and the French position on explainability of administrative algorithmic decisions as an entry point. The ethical and legal issues highlighted under the section raises challenges and broader questions regarding regulating AI and developing liability models.

Regulating Artificial Intelligence

The development of AI requires huge investment and research ecosystem, and private sector dominates the market of AI. Outsourcing the development, operation and deployment of AI models and products can create a weakening of the state control over technologies it aims to use for its citizens. As large capital flows accumulate at technology corporations whose sole agenda is to disrupt and displace innovation in order to capitalize on its market position, its leaders are appropriated in the mainstream to be change agents of established democracies thereby leading to greater concentration of wealth and influence in its corporate form (Ruggie, 2018). This is not a good development as corporations see their actions through the inherent legal flexibility of separate personality and limited liability, thereby leaving minuscule possibility to lift the veil when it comes to companies not willing to reveal algorithms even if a court were to order to do so and creates legal roadblocks for third parties to raise concerns (Hin-Yan, 2018a, 2018b).[1] As Hin-Yan Liu argues, these technical and formal barriers may lead to injustice insofar as harms and damages remain unrecognized as legal wrongs or injuries. The notion of wrong implies an infraction with potentially moral and possibly legal consequences that begins to assert the need for accountability and may occur as a direct consequence of third-parties being systematically disadvantaged by the algorithmic preferences being unable to review or challenge those veiled policies. Taken together, both the opportunities to influence the behaviour and the ability to hold the conduct to account become severely curtailed.

To monitor this 'development,' corporations must be pressed to divulge non-financial disclosures that reflect transparency and meaningful accountability for all actors. But then, how do you regulate a business that you don't understand (Milano, 2019)? Risks that AI systems will pose in the future are extremely hard to characterize and to make things less complicated, ethicists, lawyers, and computer scientists are trying to develop a normative and practical framework for trustworthy AI. For society to benefit from advances in AI technology, it will be necessary to develop regulatory policies which will manage the risk and liability of deploying systems with increasingly autonomous capabilities. If we want to allow AIs and robots to roam the internet and the physical world and take actions that are unsupervised by humans, we must be able to manage the liability for the harms they might cause to individuals and property. Resolving this issue will require untangling a set of theoretical and philosophical issues surrounding causation, intention, agency, responsibility, culpability and compensation, and distinguishing different varieties of agency, such as causal, legal and moral. Unfortunately, an extensive and illuminating scholarship is still a grey area due to lack of limited precedents for a concrete law to take shape.

Jack Balkin argues that as digital capitalism pervades every facet of life through maximum interconnectedness, locating trust and confidence of information as a fiduciary relationship becomes crucial as data becomes the most ubiquitous predictive analysis of human behaviour. As many online service providers and cloud companies should be considered as information fiduciaries with respect to their customers, clients and end-users are given that their operations are secret, and end-users do not understand and cannot monitor how their information will be used in the future makes the fiduciary obligations seem superfluous (Balkin, 2016).

In an algorithmic society, privately owned digital infrastructure and social media companies must exercise duties of good faith as social obligations and resist manipulation and control of big data that is collected, often voluntarily without explaining the possible uses of data monetization. During a disaster, it becomes pertinent that social media companies such as Facebook and Twitter should be directed to identify and surveil fake news stories and producers, block links to fake news stories and fake news sites or else supplement them with clarifying and counteracting material in order to avert mass confusion and suspicion thereby disrupting relief efforts, arresting free transportation and people movement (Balkin, 2017).

AI Liability Frameworks

Before interrogating the liability frameworks of AI, that is, if an AI-based product or model were to cause an accident or inadvertent incident, the scale of blaming can be unending as it is not a natural person. It is here that developing some form of algorithmic social contract or a robust legal personality gains currency. However, as Paul Christiano argues, meaningful human control can ensure that each decision ultimately reflects the desires of a human operator, with AI systems merely providing capabilities and advice. Unfortunately, as AI becomes more capable, such control becomes increasingly limiting and expensive. Therefore, to be safe, we must ensure that our AI systems do not cause harm by incorrectly predicting the human operator; to be efficient and flexible, we must enable the human operator to provide meaningful oversight in domains that are too complex for them to reason about unaided (Amodei et al., 2016).

Vincent Conitzer argues that most contemporary AI systems base their decisions solely on consequences, whereas humans also consider other morally relevant factors, including rights (such as privacy), roles (such as in families), past actions (such as promises), motives and intentions and biases, partiality, or lack of attention to relevant factors (Conitzer, 2019). Given that there is no codified set of ethical laws as universally recognized, it is difficult to make AI as disciplined and cautious as that of humans, argumentatively speaking. Peter Asaro argues that as profiling of humans, living conditions and geographies become available, prejudice is amplified and exacerbated by concerns over the implicit biases contained in historic data sets that get embedded into AI systems, and the obvious implications for racial, gendered, ethnic, religious, class, age, disability, and other forms of discriminatory policing, as well as how it shapes the psychology and behaviour will seemingly appear apparent. While data-driven AI techniques could have many socially beneficial applications, actually realizing those benefits requires careful consideration of how systems are embedded in, and shape, existing practices. Absent such consideration, most applications are likely to have unjust, prejudicial and discriminatory consequences (Asaro, 2012).

How to reduce this blame opacity when AI leapfrogs across automation and autonomous systems? Sven Nyholm (Nyholm, 2018) argues that irrespective of our role in using the AI systems, we are part of the 'responsibility networks' and therefore form a collaborative agency to easily navigate and locate the liability of an AI system. As the probability of miscalculations and wrong estimates surge during disasters which are ongoing or sudden such as terrorist attacks, tsunami, massive earthquakes, nuclear accidents and disease outbreaks, any harm caused by an AI system in relief delivery, targeting beneficiaries and resource allocation can make such exercises detrimental to the principal cause of saving people and property from further damage. Creating a well-meaning collaborative agency framework across AI systems and actors involved in its deployment being government agencies, local groups and administration personnel can create shared ownership of data and order control while making concrete efforts to minimize risk and possible harm that can occur due to bias in the primary data of the algorithms that make such tasks vitiated. In this context, international disaster management legal framework with its emphasis on joint responsibility could be useful.

International Legal Framework on Disaster Management: Environmental and Application of AI

A look at the international law framework regarding disaster management reveals the gradual focus of the world community to disasters starting with the United Nations declaration of the International Decade (1990–2000) for Natural Disaster Reduction (IDNDR) (Jolly, 2017). This was followed by the adoption of the Hyogo Framework for Action in 2005 for developing a resilient society which changed the paradigm of disaster management from relief and reconstruction to prevention (UN, 2005). The fundamental principles guiding the disaster laws are solidarity, joint responsibility, non-discrimination, humanity, prevention, impartiality, information, and dignity (Prieur, 2009). Disaster management legal framework emphasis on the joint responsibility of agencies and multiple actors. The point is that international disaster management law already contains provisions for collaborative and shared responsibility. The need is to create a comprehensive dialogue on integrating the application of AI and providing for a coordinated legal instrument dealing with issues of shared responsibility.

Since many of the disasters are closely linked to environmental hazards, it is imperative to look at international environmental law principles. The environmental law is premised on the principle of duty to care and precaution which has emerged as a fundamental principle of law as evidenced from judicial pronouncements and academic statements (Shelton & Kiss, 2007; Beder, 2013). At the core, environmental laws attempt to facilitate or support actions that either directly or indirectly helps reduce the intensity and frequency of hazards and vulnerabilities, and in improving disaster emergency management and recovery process. In this scenario, one of the fundamental principles often resorted by environmental decision making is the precautionary principle. The precautionary principle prescribes that scientific uncertainty should not be used as a reason to postpone cost-effective preventive measures (Kriebel, et al., 2001).

Policies based on the precautionary principle can impact AI in several ways. Firstly, the application of precaution in AI may fail to strike a balance between addressing the actual harms and the mere fear of harm. This will have the effect of distracting the policy from concentrating on the more important areas of legitimate concerns (Castro & McLaughlin, 2019). Secondly, the invocation of the precautionary principle reverses the burden of proof and places it on the agency which relies on the application of AI to prove that the activities of AI will not cause any harms. Further, any control or restrictions imposed based on precaution may limit the use and make the development of AI expensive. European Union which relies heavily on the precautionary principle in environment governance had emphasized on the incorporation of the precautionary principle in the research and commercialization of robotics. The challenge will be to find a balancing application of precaution which will not limit the innovation and address the ethical and legal concerns at the same time. If the integration of AI and the precautionary principle is based on the premise of caution, then AI through mapping poverty, disasters and engagement in service deliveries can be instrumental in advancing the goals of sustainable development goals with its overriding theme of leaving no one behind (Duberry, 2019). The key factor will be is how international agencies are setting standards and principles around the use of big data and AI based on respecting, protecting and preserving human dignity.

In this context the initiative of the European Union in developing an ethical framework regulating AI is significant. European Commission for the Efficiency of Justice (CEPEJ) has adopted certain ethical principles on the use of AI (CEPEJ, 2018). The core principles identified by the commission are as follows:

- Respect for fundamental rights
- Principles of non-discrimination
- Quality
- Transparency, impartiality, and fairness
- Access to use control

Underlining these principles is the broader public good while protecting individual interests. In addition, the European Commission which released core ethical guidelines for AI focused on the need for human agency and oversight, safety, privacy concerns, transparency, non-discrimination and fairness, societal and environmental well-being and accountability (EC, 2018). Similar concerns have been highlighted by the task force set up by the Government of India on the development of AI (GOI, 2018). These preliminary legal responses reveal a slow emerging consensus on the contours of ethical and legal concerns. The need is to engage in concrete dialogue and engagement with the broad sections of the society to concretize the contours and structure of the legal framework.

Conclusion

Application of AI has the potential to revolutionize disaster management in plausible ways. However, the development and application of AI raise fundamental legal concerns. The question arises as to how should the data access be promoted and how do we guard against biasness in the algorithms? The problems of transparency and liability need explanation especially considering the differential impact of disasters on the marginalized sections of the society. The challenge will be to design AI which will not contribute to or enhance the sufferings of the poor and marginalized.

The core point is that any development and employment of AI should keep the fundamental point of difference between a human-made decision and AI-based decision. The primary difference being the consciousness, motivation and ability to go beyond the regulatory approaches by the humans when needed. AI can only perform what has been taught and embedded through the system and cannot come up with an idea in an emergency. This distinction is paramount, and if AI is used as a tool to augment and supplement human capability, it should be embraced, while also providing for the protection of essential human values.

Note

1. As it happened in 2016 in the case of *Wisconsin v. Loomis* (881 N.W.2d 749 (Wis. Sup. Ct. 2016) where a six-year prison sentence was based in part upon the report generated by a secret algorithm. As it was a private company's proprietary software, the petition was based upon the inability of the defendant to inspect or challenge this process given that the algorithm was accused of racial bias in awarding prison sentences. The company refused to reveal the algorithm to the court citing IPR protection.

References

Aayog, N. (2018, June). *NITI Aayog*. Retrieved January 23, 2019, from NITI Aayog: http://niti.gov.in/writereaddata/files/document_publication/NationalStrategy-for-AI-Discussion-Paper.pdf

Amodei, D., Olah, C., Steinhardt, J., Christiano, P., Schulman, J., & Mane, D. (2016). Concrete Problems in AI safety. arXiv:1606.06565. Cornell University.

Aneesh, A. (2009). Global Labor: Algocratic Modes of Organization. *Sociological Theory, 27*(4), 347–370.

Asaro, P. (2012). On Banning Autonomous Lethal Systems: Human Rights, Automation and the Dehumanizing of Lethal Decision-Making. *International Review of the Red Cross, 94*(886), 687–709.

Balkin, J. M. (2016). Information Fiduciaries, and the First Amendment. *UC Davis Law Review, 49*(4), 1185–1221.

Balkin, J. M. (2017). Free Speech in the Algorithmic Society: Big Data, Private Governance, and New School Speech Regulation. *UC Davis Law Review, 51*, 1187–1221.

Beder, S. (2013). *Environmental Principles and Policies An Interdisciplinary Introduction*. London, United Kingdon: Routledge.

Blodgett-Ford, S. J. (2018). Future Privacy: A Real Right to Privacy for Artificial Intelligence. In W. B. Pagallo (Ed.), *Research Handbook on the Law of Artificial Intelligence* (pp. 307–352). Cheltenham, UK: Edward Elgar Publishing.

Bostrom, N. (2014). *Superintelligence: Paths, Dangers, Strategies*. New York: Oxford University Press.

Castro, D., & McLaughlin, M. (2019). *Ten Ways the Precautionary Principle Undermines Progress in Artificial Intelligence*. Information Technology and Innovation Foundation.

CEPEJ. (2018, December 3–4). *Council of Europe*. Retrieved January 25, 2019, from Council of Europe: https://rm.coe.int/ethical-charter-en-for-publication-4-december-2018/16808f699c

Chopra, S., & White, L. F. (2011). *A Legal Theory for Autonomous Artificial Agents*. Ann Arbor, MI: University of Michigan Press.

Conitzer, V. (2019). *Designing Preferences, Beliefs, and Identities for Artificial Intelligence*. In Proceedings of the Thirty-Third AAAI Conference on Artificial Intelligence.

Corea, F. (2018). *An Introduction to Data: Everything You Need to Know About AI, Big Data and Data Science*. Cham, The Switzerland: Springer.

Duberry, J. (2019). *Global Environmental Governance in the Information Age: Civil Society Organizations and Digital Media*. Milton, Canada: Routledge.

EC. (2018). *European Commission*. Retrieved February 10, 2019, from European Commission: https://ec.europa.eu/digital-single-market/en/high-level-expert-group-artificial-intelligence

Gao, H., Barbier, G., & Goolsby, R. (2011). Harnessing the Crowdsourcing Power of Social Media for Disaster Relief. *IEEE Intelligent Systems, 26*, 10–14.

Garnacho, A. R., Ferrers, A. I.-T., & Solanas, A. A. (2009). An Overview of Information Security. In A. Solanas & A. Martínez-Ballesté (Eds.), *Advances in Artificial Intelligence for Privacy Protection and Security, Series: Intelligent Information Systems* (Vol. 1, pp. 29–60). Hackensack, NJ: World Scientific Publishing Company.

GOI. (2018). *Report of the Task Force on Artificial Intelligence, Government of India 2018.* Department of Industrial Policy and Promotion. Government of India.

Guay, J. (2015, May 1). *Brookings*. Retrieved 5 February 2019, from Brookings: https://www.brookings.edu/blog/techtank/2015/05/01/saving-nepal-the-information-revolution/

Harris, S. U. (2018). *Participatory Media in Environmental Communication: Engaging Communities in the Periphery* (1st ed.). London: Routledge.

Hin-Yan, L. (2018a). Three Types of Structural Discrimination Introduced by Autonomous Vehicles. *UC Davis Law Review Online, 51*, 149–180.

Hin-Yan, L. (2018b). The Power Structure of Artificial Intelligence. *Law, Innovation, And Technology, 10*(2), 197–229.

HKS. (2018). *Human Rights, Ethics & Artificial Intelligence: Challenges for the next 70 years of the Universal Declaration of Human Rights.* Harvard Kennedy School.

Human Rights Big Data Development Project. (2018). The Universal Declaration of Human Rights at 70 Putting Human Rights at the Heart of the Design, Development and Deployment of Artificial Intelligence. United Kingdom: University of Essex.

IPCC. (2012). *Managing the Risks of Extreme Events and Disasters to Advance Climate Change Adaptation. Special Report of the Intergovernmental Panel on Climate Change* (p. 582). New York: Cambridge University Press.

International Telecommunication Union. (ITU). (2019). *Disruptive Technologies and their Use in Disaster Risk Reduction and Management 2019.* Geneva, Switzerland: ITU.

Jolly, S. (2017). Interrogating the Pedagogy of State Responsibility and Individual Rights in Disaster law. In A. Singh, *The Emerging Threshold of Disaster Law in Asia* (pp. 143–158). London, United Kingdom: Routledge.

Kriebel, D., Tickner, J., Epstein, P., Lemons, J., Levins, R., Loechler, E. L., et al. (2001). The Precautionary Principle in Environmental Science. *Environmental Health Perspectives, 109*(9), 871–876.

Meier, P. (2015). *Digital Humanitarians: How Big Data Is Changing the Face of Humanitarian Response* (1st ed.). Boca Raton, FL: CRC Press.

Milano, B. (2019, February 7). *Brett Milano, Big Tech's Power Growing at Runaway Speed*. Retrieved 1 March 2019, from The Harvard Gazette: https://news.harvard.edu/gazette/story/2019/02/government-cant-keep-up-with-technologys-growth/

Nyholm, S. (2018). Attributing Agency to Automated Systems: Reflections on Human-Robot Collaborations and Responsibility-Loci. *Science and Engineering Ethics, 24*(4), 1201–1219.

Pagallo, U., Corrales, M., Fenwick, M., & Forgo, N. (2018). The Rise of Robotics & AI: Technological Advances and Normative Dilemmas. In M. Corrales, M. Fenwick, & N. Forgó (Eds.), *Robotics, AI and the Future of Law* (1st ed., pp. 1–13). Singapore: Springer.

Prieur, M. (2009). *Ethical Principles on Disaster Risk Reduction and People's Resilience, Council of Europe, Executive Secretariat of the European and Mediterranean Major Hazards Agreement (EUR-OPA)*. Strasbourg, France: Council of Europe.

Ruggie, J. G. (2018). Multinationals as Global Institution: Power, Authority and Relative Autonomy. *Regulation & Governance, 12*, 317–333.

Schmitt, N. M. (2013). *Tallinn Manual on The International Law Applicable to Cyber Warfare*. Cambridge, UK: Cambridge University Press.

Series, F. I. (2018). *Harnessing Artificial Intelligence for the Earth*. World Economic Forum System Initiative on Shaping the Future of Environment and Natural Resource Security in partnership with PwC and the Stanford Woods Institute for the Environment.

Shaw, J. (2019, January–February). *Harvard Magazine*. Retrieved March 10, 2019, from Harvard Magazine: https://www.harvardmagazine.com/2019/01/artificial-intelligence-limitations#

Shelton, D. L., & Kiss, A. (2007). *Guide to International Environmental Law*. Boston: Martinus Nijhoff Publishers.

Shephered, M. (2018, January 9). *Forbes*. Retrieved 20 March 2019, from Forbes: https://www.forbes.com/sites/marshallshepherd/2018/01/09/cost-of-weather-climate-disasters-was-306-billion-in-2017-what-could-you-buy-with-that/#510e5c6f71ed

Simon, T., Goldberg, A., & Adinia, B. (2015). Socializing in Emergencies-A Review of the Use of Social Media in Emergencies. *International Journal of Information Management, 35*, 609–619.

Singh, P. J. (2018, September 20). *The Hindu*. Retrieved January 20, 2019, from The Hindu. https://www.thehindu.com/opinion/op-ed/bringing-data-under-the-rule-of-law/article24988755.ece

Sommario, E., & Venier, S. (2018). Human Rights Law and Disaster Risk Reduction. *Questions of International Law, Zoom-in, 49*, 29–47.

Stilgoea, J., Owen, R., & Macnaghten, P. (2013). Developing a Framework for Responsible Innovation. *Research Policy, 42*, 1568–1580.

United Nations. (2005). *Hyogo Framework for Action 2005–2015 Building the Resilience of Nations and Communities to Disasters.*

Verchick, R. M. (2016). Disaster Law and Climate Change. In D. A. Farber & M. P (Eds.), *Climate Change Law* (pp. 674–675). Cheltenham, UK/ Northampton, UK: Edward Elgar Publishing.

West, M. D. (2018). *The Future of Work: Robots, AI, and Automation.* Washington, DC: Brookings Institution Press.

CHAPTER 11

Artificial Intelligence and Disaster Management in Sri Lanka: Problems and Prospects

R. Lalitha S. Fernando

Background of the Study and Problem Identification

Natural disasters are frequent occurrences in Sri Lanka. Several examples are highlighted in the subsequent section to signify this phenomenon.

The tsunami in 2004 was the biggest destruction that killed 35,000 people, where 40% of deaths included children and left 900,000 homeless (Great Tsunami of December 2004 in Sri Lanka: Damage, Eyewitness accounts and Rebuilding the Economy, 2012). This is really a considerable amount in Sri Lanka, as the population is only 19 million people, with the majority of them living near the coastal area. It was Sri Lanka's worst natural disaster recorded in the history.

According to the Hong Kong Red Cross (2011), in November 2010, heavy rains throughout Sri Lanka has caused widespread flooding and

R. L. S. Fernando (✉)
Department of Public Administration, University of Sri Jayewardenepura, Nugegoda, Sri Lanka
e-mail: rlsf@sjp.ac.lk

T. V. V. Kumar, K. Sud (eds.), *AI and Robotics in Disaster Studies*, Disaster Research and Management Series on the Global South,
https://doi.org/10.1007/978-981-15-4291-6_11

threats of landslides affecting 24 out of the 25 districts in Sri Lanka. Further, they mention that the continuous rain has caused hundreds of small and medium reservoirs in the Eastern and Central parts of the country to runoff causing roads and railways to be underwater and the districts of Batticaloa and Ampara had been the worst hit with access to several areas cut off as they remain underwater.

Landslides are frequent occurrences in districts like Kandy, Nuwara Eliya and Badulla. According to the Disaster Management Centre (2011), 18 people were dead, over 52,000 families (approximately 200,000 people) were displaced in 493 temporary shelters and over 12,000 houses had been completely damaged. According to Sri Lanka's Centre for National Operations (2005) it was the second hardest-hit by natural disaster with 30,957 deaths, with another 5637 people listed as missing.

It experienced major floods in May 2016, when Sri Lanka was hit by a severe tropical storm that caused widespread flooding and landslides in 22 districts out of 25 districts in the country, destroying homes and submerging entire villages. It also reports that at least 104 people are known to have died following this disaster; 99 people are still missing, the majority due to a landslide in Aranayake, Kegalle District, which devastated three villages (United Nations Office for the Coordination of Humanitarian Affairs, 2016). Ministry of Disaster Management (2017b) reports that an estimated 301,602 people have been affected by this disaster, including at least 21,484 people who remain displaced from their homes and 623 houses have been completely destroyed and 4414 homes have been damaged. However, given that many affected locations remain underwater and others too dangerous to access due to the possibility of further landslides, it is likely that this number will rise once further assessments have been completed. On 25th May, the Government of Sri Lanka estimated that a total of 128,000 houses could have been impacted by the disaster, with 30,000 in need of reconstruction or rehabilitation (UN Office for the Coordination of Humanitarian Affairs, 2016).

In addition, Sri Lanka had been experiencing ongoing drought since the beginning of 2016. Ministry of Disaster Management (2017b) states that, approximately 1 million people had been affected across the country by the drought and paddy cultivation in the Maha season (the main harvest season) was seriously affected by the dry spell, with an expected reduction of 63% in the March/April 2017 harvest resulting in the worst main agricultural harvest season in 40 years. The Yala (minor season from May to August 2017) cultivation is also at a high risk. As a result, household food security and nutrition status were expected to deteriorate in the coming months (World Food Programme Government of Sri Lanka, 2017).

According to the Disaster Management Centre, as cited by Colombo page (2018), more than 45,000 people from nearly 14,000 families in five districts of the Northern Province had been affected due to heavy rains and floods on 22nd, December in 2018.

The above-mentioned cases are some evidences for the endless disasters occurring frequently in different facets all over the country. The occurrence of disasters cannot be prevented, but its large-scale destruction to lives and the economy have to be minimized with the use of an appropriate mechanism. Without applying a suitable mechanism to control them, valuable human lives and properties will be lost to the nation. In this context, this paper highlights the importance of applications of AI in Disaster Management.

Literature Review

Disaster is impossible to be defined by using a single word or a phrase. United Nations (2016) defines a disaster as a serious disruption of the functioning of a community or a society and disasters involve a widespread human, material, an economic or environmental impact, which exceeds the ability of the affected community to cope using its own resources. Both man-made and natural disasters are frequent occurrences which produce huge negative impact to life, property and destroy the economic, social and cultural life of people. There is no country that can prevent disasters.

Disaster management is becoming one of the focused and in dispensary fields in the development arena today since disasters are occurring at unprecedented scale (Ainuddin, 2012). The International Federation of Red Cross and Red Crescent Societies (2018) define disaster management as the organization and management of resources and responsibilities for dealing with all humanitarian aspects of emergencies, in particular preparedness, response and recovery in order to lessen the impact of disasters.

Disaster management becomes more complicated due to hazard analysis, vulnerability analysis and risk assessment. According to Denis hazard analysis focuses on the hazard types and locations, features of hazards, the environmental characteristics, which may increase the risk and vulnerability analysis concentrates on the elements at risk, which takes into consideration the possible damages, the factors which would lead to disasters, the way it will be affected, locations, durations and the capability and capacity to alter the degree of vulnerability. Risk analysis focuses on the suitability to the locals, available methods and techniques to mitigate hazards, the reliability and precision of related prediction tools. Therefore, disaster risk analysis seems to be more complicated which requires lot of data analysis. In order to manage disasters effectively, sophisticated and developed technologies have become very important.

Applicability of those technologies in Sri Lanka may not be developed and/or accessible than that of other developed countries. Based on disaster type, area that disaster occurs, applying technology or its accessibility could differ. When observing rising trends of natural disasters, disaster risk management needs to be adopted as a multidimensional endeavor, including the use of ICT (Ha, Fernando, & Mahajan, 2019).

As per Asanka, Fernando, Adhikari, Vithana Pathirage, and Karunananda (2014), Sri Lanka is also becoming a major software outsourcing country in the world next to India in the Asian region. Sri Lankan software industry consists of lots of companies from large scale to small scale. Due to the popularity of the IT industry, Sri Lankan education of IT is also strengthening and developing into an advanced level and the majority of the services offered by IT outsourcing companies are software development and product services (Asanka et al., 2014). Therefore, applying Artificial Intelligence may be much productive to be utilized in disaster management and other areas in Sri Lanka.

In 2019, Perera (2019), a top-class business leader made a compelling case for why Sri Lanka should urgently prepare a national strategy for Artificial Intelligence as a progressive move to address multiple socio-economic problems. He highlighted that Artificial Intelligence is critical for solutions to a multitude of socio-economic problems and failure will make Sri Lanka a laggard and miss benefits of this pervasive technology. Not limiting to disaster management, cost-effective AI solutions can be utilized to resolve issues in healthcare, including AI medical centers, public transportation, education, crime, security, law and order, waste management, poverty reduction and human capital development as well (Perera, 2019).

Today, AI-driven solutions are globally used in multiple sectors, while even ordinary people benefit from AI-embedded features in widely-used smartphones. "AI is not complex or overwhelming, but once you understand it properly the possibilities of solving problems with precision are endless" (Perera, 2019). Therefore, it would be more effective in applying AI in disaster management.

Artificial Intelligence in Disaster Management

Merriam-Webster defines Artificial Intelligence as a branch of computer science dealing with the simulation of intelligent behavior in computers or the capability of a machine to imitate intelligent human behavior. Governments and agencies are struggling to coordinate effective disaster relief programs with the application of AI technologies. Machine Learning (ML) is an application of AI that provides systems the ability to automatically learn and improve from experience without being explicitly programmed and it focuses on the development of computer programs that can access data and use it for self-learning (Expert System, 2019). Natural Language Processing (NLP) is also a sub-field of AI that is focused on enabling computers to understand and process human languages, to get computers closer to a human-level understanding of language (Towards Data Science, 2019). Artificial intelligence (AI), Machine Learning (ML), and Natural Language Processing (NLP) could help in predicting and related disaster management activities. Lexalytics (2019) shows that predictive analytics programs like these are still in their early stages, but offer a promising new approach to disaster relief and it is expected to apply using social data analytics and AI to respond faster to floods and other natural disasters. Further Lexalytics (2019) suggests that, by combining data from disaster response organizations, the Global Flood Detection System (GFDS), satellite flood signal, and flood-related Twitter activity, disaster relief organizations can gain a quicker understanding of the location, the timing, as well as the causes and impacts of floods. Of course, the sheer volume of Twitter data creates huge problems for anyone trying to make use of it.

Lexalytics (2019) further pointed out that in order to solve this issue; some projects are turning toward using AI and machine learning. AIDR (Artificial Intelligence for Digital Response) is an open source software platform built to filter and classify social media messages related to emergencies, disasters, and humanitarian crisis and uses supervised machine learning and artificial intelligence to tag thousands of social media messages per minute (Lexalytics, 2019). Further, it mentions that this structured data is then ready for use in dashboards, maps, or other analytics programs. Some of the most useful data generated during a crisis comes from social media users and on-the-ground aid workers; and then images and comments from Twitter, Facebook, Instagram, and YouTube. For example, these can help experts to make initial damage assessments and also can help rescue workers to find disaster victims more quickly, while identifying and mapping new disaster sites in need of aid.

Mishra et al. (2013) pointed out the importance of Robotics technology as it has become very popular in all fields of human lives. Also, they showed that an important aspect of robotics security systems is surveillance of the specified area. It would be more beneficial to have a robot during disaster conditions like earthquakes or bomb blasts, where we have to identify human beings as quickly as possible to save lives. Mishra et al. (2013) again mention that in such cases, rescue workers must gather the location information and status of victims and the stability of the structures as quickly as possible so that medics and firefighters can enter the disaster area and save victims. Normally, even though the situation is dangerous, these tasks have to be performed mostly by humans and trained dogs. There, mobile robots can help them to perform tasks that neither human, nor the dogs or existing tools can do. Also in post disaster waste management and landmine clearance, robotic systems can bring efficiency and safety to the process (Mishra, Jananidurga, Siva, Aarthi, & Komal, 2013).

Application of AI in Disaster Management

In this section several examples will be discussed in the world context.

Flood Prediction Model in Malaysia

Shahrir (2018) noted that in Malaysia, natural flood disasters frequently happen during monsoon season and Kuala Kangsar, Perak, is one of the cities with the frequent record of natural flood disasters. The author pointed out that the authority in charge of the flood disaster in Kuala Kangsar depends on the real-time monitoring from the hydrological sensor located at several stations along the main river and they failed to provide early notifications and warnings to the public even though many hydrological sensors are available. So, a flood prediction model was developed using artificial intelligence to predict the incoming flood in the area based on Artificial Neural Network (ANN) (Shahrir, 2018). ANN as the foundations of Artificial Intelligence, are the pieces of a computing system designed to simulate the way the human brain analyzes and processes information and solve problems that would prove impossible or difficult by human or statistical standards (Investopedia, 2018). As ANN has self-learning capabilities that enable them to produce better results when more data becomes available, it is expected to predict the incoming flood disaster by using information from the variety of hydrological sensors (Shahrir, 2018). This model of flood prediction can be more influential in future flood prediction not only in Kuala Kangsar, Perak but also in any areas where it is prone to floods.

Machine Learning in the USA to Be Prevented from Earthquakes

In the USA, they more focused on Machine learning to deal with disasters like earthquakes (United States Geological Survey, n.d.). With a growing wealth of seismic data and computing power at their disposal, seismologists are increasingly turning to a discipline called machine learning to better understand and predict complicated patterns in earthquake activities (Seismological Society of America, 2019).

Researchers describe how they are using machine learning methods to improve predictions of seismic activity, identify earthquake centers, characterize different types of seismic waves and distinguish seismic activity from other kinds of ground noise (Phys.org, n.d.). United Nations Office for Disaster Risk Reduction (UNDRR) (n.d.) showed once that Machine learning techniques could be used increasingly in the near future to preserve analog records of past earthquakes. According to Kaiwen Wang and colleagues, (2018) of Stanford University machine learning methods that can identify and categorize images can be used to capture these data in a cost-effective manner. Qingkai Kong (2018) of the University of California, Berkeley and colleagues pointed that machine learning methods also are already in place in applications such as MyShake, to harvest and analyze data from the crowd sourced global smartphone seismic network. According to Seismological Society of America (2019), other researchers use machine learning algorithms to filter through seismic data to better identify earthquake aftershocks, volcanic seismic activity and to monitor the earthquake that marks deformation at plate boundaries where megathrust earthquakes might occur while some studies use machine learning techniques to locate earthquake origins and to distinguish small earthquakes from other seismic noise in the environment.

Flood Alert System in India

Another technology used in India was an alert system on behalf of floods. Google is mounting up its learning from a pilot project in Patna to provide flood alerts in simple text format to people in many parts of the country using sophisticated machine learning technique and the Union Ministry of Water Resources provides Google with data on river water levels for preparing such public alerts (Sasi, 2018).

Sasi (2018) stated that in the pilot study, implemented in partnership with the Central Water Commission in India, Google showed, via public alerts, a map that included areas designated as 'high risk' 'medium risk' and 'low risk'. That pilot study used an operational hydro-dynamic model, with the explicit goal of preparing the ground for integrating Machine Learning (ML) models into the process and alerts were then sent out to individuals in the catchment areas in the form of maps and Android notifications (Sasi, 2018).

High-Performance Computing in Japan to Be Prevented from Earthquakes

UNDRR (n.d.) states that a team of researchers from the Earthquake Research Institute, Department of Civil Engineering and Information Technology Center at the University of Tokyo, and the RIKEN Center for Computational Science and RIKEN Center for Advanced Intelligence Project in Japan were qualifiers for the popular Gordon Bell Prize for outstanding achievements in high-performance computing. Earthquakes are a huge problem in many places around the world including, Japan and they use coding skills with the power of supercomputers to generate models for disaster mitigation and response (Ichimura, 2018). Realistic earthquake simulations are difficult due to wide-ranging physical phenomena operating at different scales and this complex problem led the team to devise novel strategies involving Artificial Intelligence (AI) to model earthquakes in urban centers with a high degree of accuracy (United Nations Office for Disaster Risk Reduction, n.d.).

Deep Machine Learning (ML) Algorithm in India as to Filter out Fake News

Nowadays, the spreading of fake news through media has become a huge problem and it may lead to other huge issues. Filtering out fake news about disasters on social media platforms such as Facebook, WhatsApp and Twitter is more important if possible. The Indian Institute of Technology Kharagpur has come up with a solution that uses AI to extract critical information from social media platforms that is difficult to obtain manually (Verma, 2018). Such information can be used to determine the authenticity of posts and also pass on data to aid rescue and relief operations. Ghosh, Mondal, Pramanik, & Bhattacharya (2018) of the Department of Computer Science & Engineering at IIT Kharagpur further showed that their solution can detect fake news and can even alert users in the time of disasters through deep machine learning algorithms.

UNDRR (n.d.) showed that these solutions would help to gather information that can help victims of disasters. Further, after estimations, it says that only 2% of tweets have relevant information on disasters while the remaining content is largely conversations, mostly sympathy for victims. This is how it works: while one tweet about the situation of victims hit by an earthquake is in English or Hindi, a computer program can read through the empathetic post and send the relevant information to relief operators nearby (United Nations Office for Disaster Risk Reduction, n.d.). This would help to gather more useful and reliable information for further actions during or after a disaster.

Satellite Technology in Disaster Management in Malaysia, Ethiopia and Kenya

Using satellite data to respond to environmental disasters in Malaysia, Ethiopia and Kenya is another technical solution derived to deal with disaster situations and challenges of providing a rapid response to environmental disasters as varied as flooding, drought, illegal logging and oil spills is the focus of two new projects in which the University of Oxford is a key partner (University of Oxford, 2017). The advantage of satellite data is that it can quickly identify small changes on the surface of the earth or sea that may be indicators of a larger problem in the making (National Ocean Service, n.d.). A new 'hole' appearing in a forest can provide evidence of illegal logging, or a slight color change in crops may show the early effects of drought (United Nations Office for Disaster Risk Reduction, n.d.). Combining all the data gathered from these images with other data sources has the potential to create powerful information for governments and other actors.

UNDRR (n.d.) assumes that accordingly the two projects, Earth and Sea Observation System (Malaysia) and Earth Observation for Flood and Drought Resilience in Ethiopia and Kenya, will have a total investment of £21 million and have attracted UK Space Agency funding of over £10 million. Both projects are directly relevant to many of the UN's Sustainable Development Goals. Also, it is hoped that successful lessons from the projects can be applied in other areas of the world and to other environmental threats in the future (United Nations Office for Disaster Risk Reduction, n.d.).

Application of AI in Disaster Risk Management in Sri Lanka: Examples

01. **Predicting Floods in North Central Province of Sri Lanka using Machine Learning and Data Mining Methods**

A hybrid model was developed for predicting the occurrence of floods in the North Central Province of Sri Lanka, using machine learning techniques with artificial neural networks. According to Thilakarathne and Premachandra (2017), that hybrid model developed combines two sub-predictive models in which the first model predicts the future weather-related measurements using time series modeling while the second model, which is a binary classification machine learning algorithm, predicts the probability of the occurrence of flood incidences in a future month using the forecasted weather values and historical flood data. The results of the study show that all probabilities predicted by the hybrid model are 91.7% match to the actual flood occurrences. Hence, the model could be adopted to predict flood occurrences of any region in Sri Lanka using historical weather patterns and flood-related data of the particular region. The predictive model developed has been published as an Application Programmable Interface on Microsoft Azure cloud, illustrating the practical usage and feasibility of machine learning techniques in developing modern intelligent applications (Thilakarathne & Premachandra, 2017).

02. A Multi-Agent Solution for Disaster Management

With the use of Agent Technology, a disaster management system has been implemented to provide proper management of resources and responsibilities for dealing with all aspects of a disastrous situation; in particular preparedness, response and recovery in order to lessen the impact of disasters (Perera & Karunananda, 2011). Accordingly, this system mainly consists of four agent swarms which are: forces swarm, aid store swarm, hospitals swarm, and information providers' swarm and each swarm consist of the number of agents whose tasks are explicitly defined. Agent interaction is the source of generating the intelligence; agent communication, coordination, and negotiation capabilities provide the ways and means of handling the complex nature of the domain (Perera & Karunananda, 2011).

03. Bridging the information gap through innovative Climate Information Products (CIPs)

Significant climate changes arise due to global warming and other man-made actions. Therefore, the importance of climate information is highly increasing. Information is more valuable resource especially, in the case of disaster situations. If reliable and accurate information are available, it mostly helps in managing disasters in advance. The main purpose of any CIP is to predict the future state of climate parameters with a specified period of lead time. Among many CIPs, short-to long-range weather forecasts are the most demanded and provided (The Island, 2017).

With the spread of information technology, profit oriented private providers of CIPs that are based on innovative web-based platforms have emerged and these platforms combine various business models for multiple information products that range from subscribed customized private weather reports to free supply of public weather information (The Island, 2017). These privately run CIP networks appear to be getting more and more popular with the spread of smart phones and tablet PCs. With ongoing advancements in ICT in areas such as Internet of Things (IOT), big data, artificial intelligence and drones revolution, prospects for private providers of CIPs are expected to increase tremendously in the future. There it can be more effectively introduced and utilized in the cases where disasters can occur.

Opportunities of Adopting AI for Disaster Management in Sri Lanka

Even though Artificial Intelligence is an emerging technology in Sri Lanka there are already training programs, university degree programs, learning programs, infrastructure development projects, associations, and experiments which are now underway in Sri Lanka for Artificial intelligence development and adoption.

For an example, one of major development in this area is Sri Lanka Association for Artificial Intelligence (SLAAI).

Sri Lanka Association for Artificial Intelligence (SLAAI) is a non-profit scientific association founded in 2000 devoted to the understanding of the mechanisms underlying thoughts and intelligent behavior and their emulation in machines and it was operated as the Artificial Intelligence Research group in Sri Lanka (AIRLK) for 3 years before its formal establishment in 2000 (SLAAI, 2018). The SLAAI's membership represents both academia and industry in the country and their primary objectives are to increase public awareness of Artificial Intelligence, improve teaching and research in AI, and also promote industry-academia partnerships in the use for AI techniques for real-world problem solving (Daily Mirror, 2018). Major SLAAI activities include the conduct of AI publicity programs, offering of short courses in areas of AI, promote research in AI, and conduct an annual AI conference and publishing of proceedings (SLAAI, 2018).

MSC/PG Diploma in Artificial Intelligence: The Department of Computational Mathematics of the Faculty of Information Technology of University of Moratuwa offers first-ever MSc in Artificial Intelligence (AI) program in Sri Lanka. Primary objective of this program is to impart the knowledge of theory and applications of modern AI techniques to devise intelligent software solutions thereby ensuring multifaceted career paths for the candidates (Study Portals, 2019).

This program has been designed to inculcate research skills into students through a set of unique course modules, research methodologies, philosophy of science and scientific communication (University of Moratuwa, MSC/PG Diploma in Artificial Intelligence, retrieved in 8th May 2019). It seems that it covers a wide range of AI modules from the basics to the more advanced technologies. The projects in this MSc exemplify the power of multi-agent systems, ontological modeling, brain-machine interfacing, and framework developments. A considerable percentage of projects in MSc in AI have a record of ensuring international publications annually (University of Moratuwa, Handbook, 2017).

There is an urgent need to improve the effectiveness of AI in Disaster Risk Management with international assistance and collaboration. Government of Sri Lanka should set policies and other developed mechanisms to expand AI in order to support disaster management.

Entrepreneurial ideas from knowledgeable and experienced sponsors can be taken into account in developing strategies for disaster management using Artificial Intelligence. They would assist in adopting various technologies from around the globe since they have strong connections among the nations.

Maturity of Software Industry Sri Lankan software industry has the required maturity to adopt AI technologies and they have already made their footprint on potential markets where they could find advanced AI research and development projects in different domains (Asanka et al., 2014).

Problems of Adopting AI for DM in Sri Lanka

Asanka et al. (2014) have shown several issues such as multifaceted intelligence, cooperation, affordability and manageability which may influence adopting AI. In addition, several other issues such as lack of awareness, knowledge and skills, specialists, infrastructure and financial resources, which badly affect adopting AI in disaster management, are discussed below.

Lack of Knowledge and IT Infrastructure

Lack of IT infrastructure is the biggest obstacle in the adoption of AI. Even though, ICT devices are available, many people do not have the sufficient level of knowledge on different usages of such devices. Today's young generation is much acknowledged on more applications in digital devices but with respect to disaster management, it needs to be improved through island wide.

AI systems, even the most basic ones, run on resource-intensive algorithms that require a decent combination of hardware and software capabilities, something that is not usually available in a typical office setup (Business.com, 2018). Further, this website have mentioned that an AI-ready infrastructure is capable of efficient data management, have enough processing power, be agile, flexible and scalable, and have the capacity to accommodate different types and volumes of data. In order to function, AI in Disaster Management, it needs to acquire new knowledge and needs to be updated regularly.

AI as Multifaceted Intelligence

As AI consists of many components including knowledge, reasoning, problem solving and learning. All such functions should be interconnected and necessary support should be given by expertise in each area. International support can also be taken when it is required. All the required resources and technologies should be combined together in order to have a productive outcome and IT experts or data itself alone cannot provide a sufficient support in applying Artificial Intelligence for such a comprehensive process.

Manageability

Information and data collection is another important matter. In the case of Sri Lanka because of social media, there is a huge tendency of spreading fake news in a moment. Therefore, maintaining a proper data collection and entering mechanism is a must. It may lead to estimate, predict and test the best results. Also, further development can be done with better-managed information.

Languages Usage

Applying AI using English language is easy since technology is also adopted from an international context. But when it is applied in the Sri Lankan context, understandability is very crucial. Therefore, AI needs to be developed using the main languages; Sinhala or Tamil; or else English literacy level of the entire country should be improved.

Controlling

Controllability or maintenance of AI should not be vested with the politicians or other interested parties. It should be goal oriented and messaging, focusing, agenda setting, planning, scheduling, and related other activities should be well controlled.

All the above explained problems and related obstacles have to be eliminated or at least minimized in order to have effective results of AI in disaster management.

Conclusion

As frequent occurrence of disaster has been a serious problem in Sri Lanka due to various kinds of reasons including global warming and all the natural and man-made actions, disaster management needs to be much more improved with the utilization of a more developed technology. Since human intelligence alone is inadequate to cope with the situations, Artificial Intelligence may not be considered and used as a substitute for Human Intelligence; rather, it is complementary to human Intelligence. Totally Artificial Intelligence can't be replaced to a human, but it can be used as a support to human intelligence. And also it is expected to ensure lives and property security of the country through adopting these technologies in Disaster Management. Advanced researches have to be developed for the AI projects which are already implemented. By analyzing them, we can find out what are the problems they encountered during the life cycle of development. In the context of Sri Lanka firstly it should be identified existing barriers to adopt AI in Sri Lanka and opportunities which may ease us to adopt them in Sri Lanka. Then it can be used for developing new systems for disaster management using AI. Since those projects have gone through the entire process of software development, like requirement gathering, design, development, quality assurance, implementation and support, Software Development industries have to be expanded further.

Large-scale companies need to consider adopting AI technology. And also required knowledge and workforce need to be developed through education and training. Professional organizations, research groups, academics, and the industry experts are required to play jointly to overcome barriers in applying AI technology in Sri Lankan software industry.

References

Ainuddin, S. (2012). Institutional Framework, Key Stakeholders and Community Preparedness for Earthquake Induced Disaster Management in Balochistan. *Disaster Prevention and Management: An International Journal, 21*(1), 22–36.

Asanka, P., Fernando, H., Adhikari, T., Vithana Pathirage, I., & Karunananda, A. (2014). State of Artificial Intelligence in Sri Lankan Software Industry. *IJIRT, 1*(8). ISSN: 2349-6002.
Business.com. (2018, April 13). *Articles.* Retrieved May 8, 2019, from Business. com: https://www.business.com/articles/fintech-startup-ai-adoption/
'Centre for National Operations for Relief Work', Daily News, 1 Sri Lanka 2005, www.dailynews.lk/2005/01/01/new02.html accessed June 2020.
Colombo Page New Desk. (2018, December 22). Heavy Floods in Northern Districts of Sri Lanka Displace Thousands, Reservoirs Overflow. Retrieved from http://www.colombopage.com/archive_18B/Dec22_1545495185CH.php
Daily Mirror. (2018, May 18). *Positive Use of AI for All – EDITORIAL.* Retrieved May 8, 2019, from Dailymirror: http://www.dailymirror.lk/article/Positive-use-of-AI-for-all-EDITORIAL-150068.html
Disaster Management Centre. (2011). *Annual Report, Sri Lanka Government* available at https://www.parliament.lk/uploads/documents/paperspresented/annual_report_disaster_management_centre_2011.pdf
Expert System. (2019, May 6). *Machine Learning.* Retrieved from https://www.expertsystem.com: https://www.expertsystem.com/machine-learning-definition/
Gosh, S., Mondal, T., Pramanik, P., & Bhattacharya, I. (2018). Analysis and Early Detection of Rumors in a Post Disaster Scenario. *Information Systems Frontiers, 20*(5), 961–979. Retrieved May 9, 2019.
Great Tsunami of December 2004 in Sri Lanka: Damage, Eyewitness accounts and Rebuilding the Economy. (2012). Retrieved from http://factsanddetails.com/
Ha, H., Fernando, L. S., & Mahajan, S. K. (2019). Disaster Risk Management in Agricultural Sector in South Asia. In H. Ha, L. S. Fernando, & S. K. Mahajan (Eds.), *Disaster Risk Management in Agriculture* (pp. 157–168). New York: Business Expert Press.
Ichimura, T. (2018, November 16). *Public Release.* Retrieved from EurekAlert: https://www.eurekalert.org/pub_releases/2018-11/uot-erf111618.php
IFRC – The International Federation of Red Cross and Red Crescent Societies. (2018). About Disaster Management. *About Disaster Management.* Retrieved from https://www.ifrc.org/en/what-we-do/disaster-management/about-disaster-management/
Investopedia. (2018, March 9). *Artificial Intelligence.* Retrieved from investopedia.com: https://www.investopedia.com/terms/a/artificial-neural-networks-ann.asp
Kong, Q., Inbal, A., Allen, R., Lv, Q., & Puder, A. (2018). Machine Learning Aspects of the MyShake Global Smartphone Seismic. *Network, 90.* https://doi.org/10.1785/0220180309

Lexalytics. (2019, May 6). *Artificial Intelligence for Disaster Relief: A Primer.* Retrieved from https://www.lexalytics.com: https://www.lexalytics.com/lexablog/artificial-intelligence-disaster-relief

Ministry of Disaster Management. (2017a). *Initial Rapid Assessment on Drought 2016/17.* Retrieved May 9, 2019, from https://www.wfp.org/sites/default/files/SLA_Drought_20170119_updated.pdf

Ministry of Disaster Management. (2017b). *Sri Lanka Rapid Post Disaster Needs Assessment Floods and Landslides.* The Ministry of Disaster Management & Ministry of National Policy and Economic Affairs in collaboration with the United Nations, World Bank and European Union. Retrieved May 9, 2019, from https://www.undp.org/content/dam/srilanka/docs/localpublications/PDNA%20Sri%20lanka%202017-1.pdf

Mishra, M., Jananidurga, P., Siva, S., Aarthi, U., & Komal, S. (2013). Application of Robotics in Disaster Management in Land. *International Journal of Scientific and Research Publications, 3*(03), 1–4.

National Ocean Service. (n.d.). *Ocean Facts.* Retrieved May 7, 2019, from National Ocean Service: https://oceanservice.noaa.gov/facts/satellites-ocean.html

Perera, D. (2019, March 19). Dhammika Perera calls for National AI Strategy. *Daily News.* Sri Lanka. Retrieved from http://dailynews.lk/2019/03/19/business/180656/dhammika-perera-calls-national-ai-strategy

Perera, P., & Karunananda, A. S. (2011, December 20). MASDM – A Multi Agent Solution for Disaster Management. *Proceeding of the eighth Annual Sessions.*

Phys.org. (n.d.). *Seismological Research Letters.* (S. S. America, Producer) Retrieved May 7, 2019, from phys.org: https://phys.org/news/2019-03-machine-characterize-earthquakes.html

Sasi, A. (2018, December 28). *ExpressTech.* Retrieved May 7, 2019, from The Indian Express: https://indianexpress.com/article/technology/tech-news-technology/google-now-ready-to-give-public-flood-alerts-5512705/

Seismological Society of America. (2019, March 1). *Sciencedaily.com.* Retrieved May 7, 2019, from Sciencedaily.com: https://www.sciencedaily.com/releases/2019/03/190301123247.htm

Shahrir, N. S. (2018). Flood Disaster Prediction Model Based on Artificial Neutral Network: A Case Study of Kuala Kangsar, Perak. *Improving Flood Management, Prediction and Monitoring.*

Sri Lanka Association for Artificial Intelligence (SLAAI). (2018). *Home.* Retrieved May 8, 2019, from Sri Lanka Association for Artificial Intelligence (SLAAI): http://slaai.lk/

Study Portals. (2019, May 8). *Studies.* Retrieved from Study Portals Masters: https://www.mastersportal.com/studies/197085/artificial-intelligence.html

The Hong Kong Red Cross. (2011, January 18). *Sri Lanka Floods and Landslides 2011.* Retrieved May 8, 2019, from Reliefweb: https://reliefweb.int/report/sri-lanka/sri-lanka-floods-and-landslides-2011

The Island. (2017, July 14). Climate Challenge: Bridging the information gap through innovative climatic information products (CIPs). Upali Newspapers (Pvt) Ltd. Retrieved from https://www.island.lk/index.php?page_cat=article-details&page=article-details&code_title=168241

Thilakarathne, H., & Premachandra, K. (2017). *Predicting Floods in North Central Province of Sri Lanka using Machine Learning and Data Mining Methods.*

Towards Data Science. (2019, May 6). *An Easy Introduction to Natural Language Processing.* Retrieved from https://towardsdatascience.com: https://towardsdatascience.com/an-easy-introduction-to-natural-language-processing-b1e2801291c1

UN Office for the Coordination of Humanitarian Affairs. (2016). *Sri Lanka: Floods and Landslides Situation Report No. 2 (as of 26 May 2016).*

United Nations Office for Disaster Risk Reduction. (2016). *Terminology.* Retrieved May 6, 2019, from https://www.unisdr.org: https://www.unisdr.org/we/inform/terminology

United Nations Office for the Coordination of Humanitarian Affairs. (2016, May 26). *ReliefWeb.* Retrieved May 6, 2019, from ReliefWeb: https://reliefweb.int/disaster/fl-2016-000050-lka

United States Geological Survey. (n.d.). *Earthquake Hazards Program.* Retrieved May 7, 2019, from earthquake.usgs.gov/: https://earthquake.usgs.gov/learn/facts.php

University of Moratuwa. (2017). *Handbook.* The External Affairs, Publicity and International Student Promotion Division of the University of Moratuwa.

University of Oxford. (2017, October 19). *News and Events.* Retrieved from ox.ac.uk: http://www.ox.ac.uk/news/science-blog/using-satellite-data-respond-environmental-disasters#

Verma, P. (2018, November 20). *Tech.* Retrieved May 7, 2019, from The Economic Time: https://economictimes.indiatimes.com/tech/software/iit-kharagpur-develops-tech-to-filter-fake-news-about-disasters/articleshow/66699717.cms

Wang, K., Ellsworth, W. L., Beroza, G. C., Williams, G., Zhang, M., Schroeder, D., et al. (2018). Seismology with Dark Data: Image-Based Processing of Analog Records Using Machine Learning for the Rangely Earthquake Control Experiment. *Seismological Research Letters.* https://doi.org/10.1785/0220180298

World Food Programme Government of Sri Lanka. (2017). Sri Lanka – Initial Rapid Assessment on Drought 2016/17, 15 January 2017.

CHAPTER 12

Applications of Artificial Intelligence in Reconstruction Governance Lessons from Nepal Earthquakes

Narendra Raj Paudel

Background

Studies show that significant number of natural hazards have been observed in Nepal, resulting in a high level of vulnerability (Aryal, 2012a, Aryal, 2014, UNDP, 2001, BCPRA, 2004, GHI, 2001). Every year Nepal faces challenges of natural hazards such as floods, fires, landslides, inundations, glacier lake outbursts, storms and earthquakes (Aryal, 2012b). These natural hazards have been causing loss of life and property of Nepalese people. Among these natural hazards, earthquake was found to be critical because it heavily jeopardizes livelihood of people, settlements, developmental infrastructures and others simultaneously (Paudel, 2016). Its coverage is comparatively high compared to other natural hazards because Nepal lies in an earthquake-prone area of Indian plate and ranks 11th in the global position of risk. On April 25 and May 12, 2015, earthquakes of 7.8 magnitude hit 31 out of 75

N. R. Paudel (✉)
Public Administration Campus, Tribhuvan University, Kirtipur, Nepal

T. V. V. Kumar, K. Sud (eds.), *AI and Robotics in Disaster Studies*,
Disaster Research and Management Series on the Global South,
https://doi.org/10.1007/978-981-15-4291-6_12

districts of Nepal badly. About 9000 people died and 17,000 people were injured. Moreover, about 4000 public buildings and 600,000 private houses were destroyed. About 136 settlements were relocated. Altogether NRP 700 billion worth property was lost due to the earthquakes (NPC, 2015). Prior to these earthquakes, Nepal was stroked by earthquakes in 1255, 1934, 1980, 1988 and 2011. Often, the root causes of increase in such disasters and heavy impacts on life and property of people include weak governance, population growth, rapid urban expansion, relatively weak land use planning, spread of informal settlements, poor construction methods, steep land farming practices, encroachment of settlements into river planes and forest areas and environmental degradation (Aryal, 2012a). A study conducted by Nepal Red Cross Society and the Practical Action Nepal (2012) shows negative impacts of disaster incidents on local livelihoods and ecosystem of Nepal (quoted from Aryal, 2012b). Paudel (2016) argues that the better the response and performance of public institutions is, the greater the trust is where disasters like earthquakes are concerned. The experience of Latin America stresses the importance of past disaster incidents' analyses for current and future development planning (Lavell and Lavell, 2009; Wilches-Chaux, 2007). Therefore, data-based planning and implementation before disaster and after disaster is crucial and challenging. In this context, it is pertinent to explore how Nepal has been managing all responses, reliefs, rehabilitations and reconstruction works successfully with the assistance of foreign countries. The volumes of reconstruction and rehabilitation works are huge because of heavy loss of life and property, destroyed developmental infrastructures and damaged livelihood of people. So, it demands more resource and more capable institutions as well as rationalized decisions for policies, programs and projects for reconstruction and rehabilitations to resume livelihood of people as before. The literature related to artificial intelligence (AI) shows that its applications to collect, store and analyze the data from computer-supported software take correct decisions during relief, response, reconstruction and rehabilitation works caused by earthquake (Ramchurn et al., 2016 and Deogade, 2018). This chapter examines the reconstruction and rehabilitation governance after 2015 earthquakes hit Nepal through the lens of the AI. More specifically, this chapter analyzes the roles of institutions created for reconstruction of damages due to earthquake hit and the processing of information for better decisions through AI lens, and brings into focus factors which hinder reconstruction governance in Nepal.

Artificial Intelligence and Reconstruction Governance in Nepal

The AI, despite its old roots, has been used in recent times for prediction and forecasting, early warning system, resilience infrastructure and resilience planning of disaster as well as reconstruction governance of earthquake. Conceptually, AI is an activity devoted to making machines intelligent and intelligence is a quality that enables an entity to function appropriately and with foresight in its environment (Nilsson, 2010). Collaborative system is one of the AI research models which helps develop autonomous systems that can work collaboratively with other systems and with humans (Stanford University, 2016). This research relies on developing formal models of collaboration and studies the capabilities needed for systems to become effective partners. There is growing interest in applications that can utilize the complementary strengths of humans and machines—for humans to help AI systems to overcome their limitations and for agents to augment human abilities and activities. By applying this form of AI model, the machine-based intelligence, human intelligence and their combinations make effective decisions for reconstruction governance in Nepal. Machine-based intelligence is about computer-based programs which work automatically. Computer analyzes the complex and uncertain data fed to the system collected through fields. For the purpose of reconstruction governance in Nepal, field information such as details of loss of life and property, and damages of public and private entities, geological information, banking information for resource allocation and expenditure, achievements of reconstruction made so far, and public grievances are very much useful for decisions regarding plans and policies. As a result, beneficiaries are correctly located and achievements of reconstructions are illustrated. Besides, the human-based intelligence fills the gaps for the tasks that cannot be automated. Human intelligence corrects and validates the machine's outputs of reconstruction governance. Combining both machine and human intelligence gives clear picture of tasks that are needed for reconstruction governance. The AI model for reconstruction governance in Nepal is presented below.

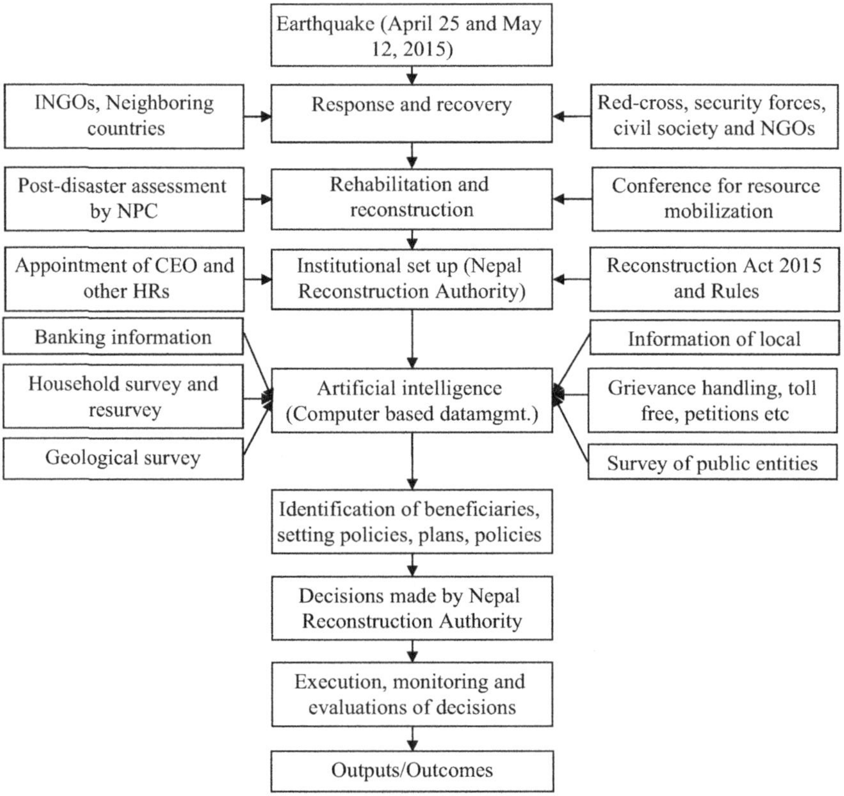

Source: Constructed by author

Research Methodology

The study is conducted sequentially. First, it analyzes contents from policies, programs and institutional structures. Second, it examines the information management system developed by Nepal Reconstruction Authority (NRA) through application of AI. The software which supports to analyze information of beneficiaries, banking, loss of life and property, damages of physical property, grievances and so on gives exact information of reconstructions of damages and progress along with benefits received by beneficiaries. Third, it also examines the decisions made by NRA and subsequent achievements made so far. In the fourth stage, issues of reconstruction governance which were identified from information system and content analysis are also validated through two case studies, one from Dhadhing district and the other from Kathmandu district. While choosing cases, one is about a success story related to private house buildings in Dhadhing district and another case related to a story of unsuccess of Ranipokhari heritage site reconstruction in Kathmandu district. To test the issues, data in case studies were collected through story narrations and interviews with five house owners, conservationist, officials from archeology department and ministry and chief district officers.

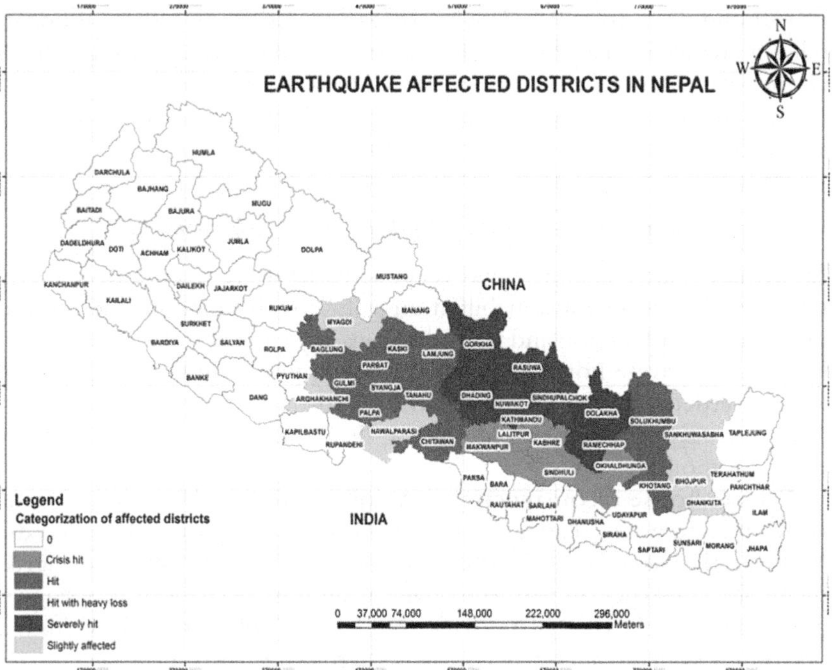

Reconstruction Governance in Nepal

Response and Relief

Immediately after the earthquakes of 2015, relief and response teams of government, private sectors, and civil society and foreign countries with relief materials were mobilized in Nepal. The information of causality and damages was collected from security forces, tweeter analyses, phone calls and media. Ministry of Home Affairs, Government of Nepal (GoN), was the core coordinating agency which established coordination centers at the ministry and Tribhuvan International Airport, Kathmandu. Even though 131,610 security personnel (65,059 staff of Nepali Army, 41,776 staff of Nepal Police and 24,775 staff of Nepal Armed Police), 22,500 civil servants, 4000 private organizations and nongovernmental organizations (NGOs) including 60 countries and international NGOs (INGOs) were deployed for the sake of relief and response, there was coordination problem among the actors. People who were in accessible places got abundant relief materials, while none were available in remote areas to support the earthquake victims. People also observed heavy dependency on donors for relief materials. They did not do what they actually could. This trend created problems for relief and response purpose.

Reconstruction and Rehabilitation Achievements

Immediately after response and relief activities were over, GoN entered into reconstruction and rehabilitation stage. In this section, the achievements in reconstruction and rehabilitations stage made so far are evaluated in terms of private house reconstruction, integrated settlement and risky zones' settlement identification, and reconstructions of public entities (schools, universities, roads, heritages, drinking water etc.) through the application of AI.

Private house reconstructions: According to initial survey and resurvey of private house damages, about 900,000 houses were damaged due to the earthquakes of 2015. Among them, 92 percent people made agreements for construction with the financial support of NRA at first stage as per data of July 2018 (NRA, 2018). NRA had a policy to disperse fund as per progress made by people so that there won't be any misuse of the fund. Only 65 percent people got second installment and 43 percent of those who completed their houses received final installment. To provide justice to the earthquake victims, NRA collected 450,000 grievances through phone, toll-free phone and other means as specified by laws. Because of the grievance handling mechanism, 130,000 new victims were reidentified as earthquake victims. Here, AI made it easy to ensure that "None would be repeated and none would be left". The private house reconstruction trends since 2015–2018 were not so satisfactory as expected because of the lack of information on actual victims, bureaucratic hassles, no availability of technical and administrative authorities as and when required for beneficiaries, no bank accounts of beneficiaries, no citizenship certificate, lack of appropriately owned land, disputes within family because of joint family and so on (interview with beneficiaries and officials, 2019). On the one hand, authorities had to comply with laws. One the other hand, people lacked required proof to get support for reconstruction. In between this process, bureaucratic attitude also hampered to expedite the process. Such mismatches created slow decision-making for private house reconstruction.

Risky zones' resettlement and integrated settlement identification: One of the objectives of Natural Disaster Act 2015 and its regulation is to promote integrated settlement and avoid risky zone for settlement for safety. Information generated through geological survey is the main basis to identify unsafe and safe settlements, and those settlements that can be made safe, and corresponding beneficiaries who lived in such places. The AI made it easy to find out 43 percent safe zones, 33 percent that can be made safe and 24 percent unsafe zones for settlement through survey of 1075 risky zone settlements. AI also identified 5000 corresponding beneficiaries who lived in risky zones. Despite additional grant of NRP 200,000 to purchase land in safe zone, only 34 percent people purchased land for reconstruction of their houses. Besides, 14 percent people were also provided land for reconstruction of their houses at safe zone. However, remaining 52 percent had to stay in such risky zones. In case of integrated settlement, only 6 percent out of 37 integrated settlements were completed within stipulated time frame even though 86 percent (29) settlements had made agreements according to the data of July 2018. Only 12 percent (six) settlements were in the process of reconstruction. Despite such activities for risky zones' resettlement and integrated settlement, the achievements are negligible. From interviews, reasons responsible for low achievements in resettlement of people include difficulties to coordinate with other ministries, technical difficulty for geological survey and its analysis, people's emotional attachment with their own parental land and so on. Such reasons led to low achievements in resettlement programs of the earthquake victims.

Public entities' reconstructions: Public entities refer to schools, university buildings, health posts, government offices, offices of security forces, heritages, roads and drinking water supply. Fifty-five percent of 7553 schools, 7 percent of 30 university buildings, 54 percent of 1197 health posts, 60 percent of 415 government offices, 28 percent of 383

Table 12.1 Public entities' renovations

Public entities	*Damaged by earthquake (N)*	*Renovated (%)*	*Ongoing (%)*	*Remaining (%)*
Schools	7553	55	25	20
Universities	30	7	70	23
Health posts	1197	54	12	34
Government offices	415	60	35	5
Security offices	383	28	28	44
Heritages	753	28	30	42
Roads	86 (1151 km)	4	69	23
Drinking water	3212	25	21	55

Source: Nepal Reconstruction Authority, Kathmandu, July 2018

security buildings, 28 percent of 753 heritages, 4 percent of 1151 km road and 25 percent of 3212 drinking water supplies were found to have been renovated as of data July 2018. In these public entities, renovating tasks were also going on. However, 20 percent schools' renovations were still found to be remaining. Likewise, reconstruction of 23 percent university buildings, 34 percent health posts, 5 percent government offices, 44 percent security forces' offices, 42 percent heritages, 23 percent road constructions and 55 percent water supply remained, even though their renovations are critically important because they make the livelihood of victims easier and better. The slow achievements in these sectors are due to slow decisions from stakeholders, difficulties to identify need of public entities and lack of appropriate manpower. The renovation faced difficulty especially in heritage sites because of their specific codes and standards to assure (interviews with officials) (Table 12.1).

Factors Affecting Reconstruction Governance in Nepal

Despite extensive effort for reconstruction through the application of AI, low achievements have so far been observed against expectations which were prescribed in NRA's strategic five-years plan (2015–2019). Now, after four years of reconstruction governance only about 40 percent of achievements have been made. Therefore, it is indeed pertinent to explore the reasons why decisions were not implemented effectively. The assumption is that high-quality decision implementation caused fast reconstruction of damages caused by the earthquakes of and vice versa. The reasons observed from field study and documents are as follows.

Institutional set up: Reconstruction of Earthquake Affected Structure Act 2015 states that the Nepal Reconstruction Authority (NRA) is the sole agency which aims at prompt completion of structure damaged due to an earthquake in a sustainable, resilient and planned manner. It also aims to promote national interest and provide social justice by ensuring resettlement and translocation of the persons and families displaced by the earthquake. To achieve these objectives, national advisory council under the leadership of prime minister, advisory councils and steering committee is formed in NRA. For running day-to-day activities, a chief executive officer (CEO) is appointed. Thus, the institutional structure of NRA is responsible for formulation of plans and policies, data/information collection from field, resource generation and its mobilization, allocation of duties and responsibilities, institutional linkages with other ministries and local governments, and monitoring and evaluation of their works. Thus, main functions of NRA are to deliver services to earthquake victims who are in both remote and accessible areas from Kathmandu. Because of the centralized structure of NRA, the decision-making power and authority for reconstruction was not devolved at field offices. They were dependent on center for taking small decisions of reconstruction despite correct options of decisions generated by AI. On the flip side, the nature of reconstruction work demands spot decisions. Besides, the officials who had been working at field level were always demanding their own benefits rather than service delivery to the needy people. Such an attitudinal problem of center and local authority of NRA made them suspicious of each other. The lengthy process of decision-making, conflict between center and local officials and mismatched structure of NRA caused slow decision-making of reconstruction as well as its implementation. Ultimately, such an institutional setup was responsible for slow reconstructions of damages due to earthquake.

Policies for reconstructions governance: Under the broad umbrella of Reconstruction of Earthquake Affected Structure Act and its Regulation 2015, more than one dozen policies (Appellate Procedure 2016, Land Acquisition Procedure 2015, Land Registration Procedure 2015, Resettlement Procedure, 2015, Private House Reconstruction Grants Disbursement Procedure, 2016, Hazardous Settlement Relocation and Reconstruction Procedure 2016, Grievance Handling Procedure related to Reconstruction, 2017, Training Operation and Management Directive

2016 etc.) were framed by the NRA. However, the implementation of such policies was found to be more uncertain and complex. For example, people who had access to authority got subsidies loan rather than the needy people. More than 50 percent of the earthquake victims registered grievances which were collected by the system. More and more grievances meant more confusion in reconstruction policy implementation. Very few people got livelihood support–training opportunity. Resettlement program was also found to be unsuccessful. Coordination with other ministries for land acquisition was also found to be problematic. Collectively, such tendencies of policy implementation also created hurdles for completion of reconstruction tasks despite such a policy having been formulated through application of AI.

Case Study 1 (Success Story of Reconstruction Governance)
Out of 31 districts hit badly by the earthquakes of, Dhadhing, which was affected the most, was chosen to study reconstruction governance in Nepal. The rationale given for the study was that Dhadhing covered comparatively biggest area among the districts. Dhadhing, except Kathmandu, has the highest population size. Besides, Dhadhing, which is extended from Himalayan region to Terai ecological region of Nepal, lies in hilly ecological region of Nepal. Therefore, the knowledge generated from Dhadhing district might be representative for other districts in terms of reconstruction governance in Nepal. The earthquake damaged 100,000 houses. About 90 percent of house owners made agreement to reconstruct their houses with support of NRA grants. As of data of July 2018, 40 percent houses were already constructed and remaining 60 houses were under construction. House owners were asked about the services provided by the NRA and difficulties they faced during their house construction. In an interview, five houses owners whose houses were completed argued that it became a bit difficult to construct their houses due to lack of water, geographical remoteness and lack of skilled masons. They also faced hurdles to manage additional funds other than government grants to construct their houses. They stated that they could not get authorized persons to monitor and supervise the sites. They were satisfied because at least now they had their own earthquake-proof houses to live in. Other people whose

(*continued*)

Case Study 1 (continued)
houses were under construction stated that they also faced same kinds of problems in addition to division of their parental property, issue of citizens certificate, recommendations from ward office of local government and issue of constructing materials. However, chief district officer of Dhadhing district argued that beneficiaries showed apathetic behavior. The NRA also had low number of technical persons to monitor and supervise the sites. People did not show their sense of ownership for this reconstruction venture even though NRA provided grants to them for their house reconstruction.

Resource mobilization: Resource mobilization is a crucial aspect of reconstruction governance. Revised post-disaster relief fund estimated that NRP 738 billion was required for reconstruction and revival of earthquake-affected people's livelihood. Initially, it was NRP 938 billion as per post-disaster relief fund for five years (2015–2019). To mobilize such fund, donor agencies made commitment for NRP 410 billion. On its contrary, donors made agreement with GoN only for NRP 262 billion. It shows that there was huge resource gap between commitments made by donors and actually they did agreement with GON. In case of expenditure, NRP 218 billion, of which NRP 151 billion were given by GoN and

NRP 67 billion by donors, was spent for the reconstruction. Thus, the data shows that huge financial resource gaps created hurdle in implementation of plans, policies and programs related to reconstruction governance in Nepal though plans, policies and programs were set properly with the application of AI.

Case Study 2 (Heritage Reconstruction: Unsuccess Story)
Ranipokhari is a century-old pond constructed by King Pratap Malla in memory of his son and queen in 1670. Initially, it was decided that the reconstruction of Ranipokhari will be led by Kathmandu Metropolitan City (KMC). KMC planned to commercialize the Ranipokhari to raise revenue for metropolitan city. KMC's decision to develop the historic pond into a recreational park was also vehemently protested by locals and conservationists. In addition, there was a tug-of-war between Department of Archeology and KMC in the design of Balgopaleshower Temple, which is situated in the middle part of Ranipokhari. The reconstruction of the Balgopaleshower Temple, situated in the premises of Ranipokhari, had been mired in controversy one more time, with the Ministry of Culture, Tourism and Civil Aviation asking the Department of Archaeology to reconsider its decision of rebuilding the historic property in a style that imitates the architecture of a Gumbaz structure. The directive from the ministry had come at a time when the expert panel, responsible for the reconstruction of the temple, had recommended the structure to be rebuilt in Garanthakut structure. It is believed that Balgopaleshower Temple was originally built in Garanthakut design by King Pratap Malla in 1670. The temple was later rebuilt in Gumbaz model following the great earthquake of 1934. Finally, NRA, ministry and archeology department and expert panel decided to build temple as per Malla era. After a three-year hiatus, and expenditure of more than NRP 25 million, reconstruction of Ranipokhari finally began on January 28, 2019. Reconstruction of Ranipokhari was delayed after locals protested the use of modern construction materials for reconstruction of the pond and design of Balgopaleshower Temple.

(*continued*)

Case Study 2 (continued)

Politicization: With the appointment of a CEO, NRA became hotbed for politics. Initially, Nepali congress (ruling party during disaster) and UML including other communist parties were quarrelling for the executive post of NRA. In the meantime, government changed and new constitution was promulgated; UML became ruling party and Nepali Congress became opposition party. Again their fighting for executive post was repeated. This trend was transferred to local bodies as well. Newly formed local bodies also did not openly support the venture of the NRA. Huge resources and huge concerns of politicians have created problems for speedy work of reconstruction. Politicians' concerns did not facilitate the process of reconstruction; they only created hurdles for reconstruction, even though they claimed they will work for the earthquake victims.

Maintaining originality of heritage sites: Due to lack of proper identified codes and standards, material used in heritage sites caused delay in their reconstruction. Archeology department also did not have their original standards. More study was needed. For example, heavy disputes and contradiction were put forward in case of Ranipokhari's reconstruction. There was a lack of skilled manpower for their reconstructions. Therefore, it caused delay in their reconstruction.

Others: People's sentiment attached to their parental land, bureaucratic attitude, coordination with other ministries, livelihood of people, resource gaps, appropriate land for resettlement, identification of beneficiaries, skilled manpower also contributed to delay in reconstruction process in Nepal to implement decisions made through AI application.

Conclusions

Reconstruction governance is based on cognition of human elements despite many codes and standards developed in the application of AI. For proper decision-making related to reconstruction governance, the NRA had collected information related to loss of life and property, damages to developmental infrastructures, banking information, geological information and achievements made so far. Information was fed into computer-based software and complex data set was analyzed. The guiding principle was "none repeated and none left for reconstruction of damages due to earthquake 2015". Inferences of such data analysis were used for decision-making by using human intelligence. Combining both computer-based intelligence and human intelligence led to proper decisions on reconstruction governance-related plans and policies. As a result, there were more than one dozen policies, about 40 percent reconstruction work was completed, and NRP 200 billion of financial resources was spent. A bit of weak automation was to challenges of validation and cross validation of data collected from field. It took a bit long time for decision-making. The reconstruction governance was in slow motion because of many unseen challenges. The major highlighted wicked problems were political transition and instability/politicization, resource gap between donors' commitment and reality, identification of beneficiaries, bureaucratic attitude, coordination with stakeholders, emotional feeling of people, lack of trained manpower and so on. These challenges were found to be responsible for hurdles faced in the completion of the reconstruction within its stipulated time frame. Thus, the reconstruction governance in Nepal through the application AI reveals that it is neither satisfactory nor unsatisfactory.

References

Aryal, K.R. (2012a). Getting Down to Local Level: Exploring Vulnerability to Improve Disaster Management System in Nepal, PhD thesis, Newcastle: Northumbria University.

Aryal, K. R. (2012b). The History of Disaster Incidents and Impacts in Nepal 1900–2005. *International Journal of Disaster Risk Science, 3*(3), 147–154. https://doi.org/10.1007/s13753-012-0015-1

Aryal, K. R. (2014). Disaster Vulnerability in Nepal. *International Journal of Disaster Risk Reduction, 9*, 137–146.

BCPRA. (2004) Global Report: Reducing Disaster Risk: A Challenge for Development, New York. Access from http://www.undp.org/cpr/whats_new/rdr_english.pdf

Deogade, S. (2018). Implementation of Data Mining and Machine Learning Techniques in the Context of Disaster and Crisis Management. *International Journal of Scientific Research in Science and Technology, 4*(8), 46–49.

GHI. (2001). *Global Earthquake Safety Initiative-Pilot Project Final Report.* Palo Alto: Geo Hazards International and United Nations Centre for Regional Development. Available at http://geohaz.org/new/images/publications/gesi-report%20prologue.pdf

Lavell, A., & Lavell, C. (2009). *Local Disaster Risk Reduction: Lessons from the Andes.* Lima: Communidad Andina.

Nilsson, N. J. (2010). *The Quest for Artificial Intelligence: A History of Ideas and Achievement.* Cambridge: Cambridge University Press.

NPC. (2015). *Post Disaster Assessment.* Kathmandu: National Planning Commission.

Paudel, N. R. (2016). Citizen's Trust in Public Institution in Nepal After April 25, 2015 Earthquake. *South Asian Journal of Policy and Governance, 40*(2), 97–116.

Ramchurn, S. D., Huynh, T. D., Wu, F., Ikuno, Y., Flann, J., Moreau, L., et al. (2016). A Disaster Response System Based on Human-Agent Collectives. *Journal of Artificial Intelligence Research, 57*, 661–708.

Stanford University. (2016). *One Hundred Years Study on Artificial Intelligence (AI100).* USA: Stanford University. Accessed August 1, 2016, https://ai100.stanford.edu

UNDP. (2001). Disaster Profile of Least Developed Countries, Third United Nations Conference on Least Developed Countries, Brussels, 14–20 May 2001. Access from http://www.undp.org.cpr/disred/document/publication/dpldc.pdf

Wilches-Chaux, G. (2007). *ENSOWHAT?: LA RED Guide to Getting Radical with Enso Risk.* Bogota: ARFO Editores e Impresores.

CHAPTER 13

ICT Infrastructure of Disaster Management in India

G. Durga Rao

Introduction

This chapter surveys the applications of modern technologies in the administration of information and communication in disaster management activities. The role of information and communication in disaster management is irrefutable. Reliable and updated information is a baseline requirement for effectively managing disasters. Information on hazards, risks and vulnerabilities of various disasters is a prerequisite for holistic disaster management. Knowledge of disaster occurrence and human, social and economic costs of disasters is helpful in analysing the temporal and geographical trends of disaster impact (Nair, 2012). Primary information pertaining to the likely impact of the disaster in terms of the extent of the area affected, community affected, availability of resources for evacuation of the people and relief and quick assessment of damages, is vital for effective disaster management (NDMA guidelines, 2012). Informed communities can make the task of disaster management a lot more effective

G. D. Rao (✉)
Central University of Jammu, Bagla, India

T. V. Kumar, K. Sud (eds.), *AI and Robotics in Disaster Studies*, Disaster Research and Management Series on the Global South,
https://doi.org/10.1007/978-981-15-4291-6_13

and systematic. It is impossible to conceive a foolproof disaster management strategy without adequate information. Individuals and institutions are often forced to make unreliable, inaccurate and conflicting decisions in the absence of proper information and communication. Communication also plays a crucial role in disaster risk reduction and disaster management. Communication helps in alerting communities, monitoring activities and assessing risks, costs and benefits associated with disaster management. According to Sunita Kuppuswamy (2010);

> Communication raises awareness of the hazards about the disaster. It provides a means of alert and early warning to the people in the disaster prone zones. It will be helpful in taking preventive measures to avert disasters. It is needed for giving training and awareness to the people of the disaster prone zones (regarding various aspects related to disasters). It also provides data for integration and analysis of spatial and temporal disaster data, modeling and simulation disasters more precisely. It helps in real-time decision making and enhance emergency response capabilities (p. 27).

With the emergence of Information and Communication Technologies (ICTs from here on wards), there has been a remarkable change in the administration of information and communication in disaster management activities. ICTs serve as tools to improve the roles and capacities of all the stakeholders while facing hazards, risks and the resultant disasters. The adverse impact of disasters can be minimized by using state-of-the-art Information and Communication Technology systems. ICTs can help exponentially in planning and implementation of disaster management activities. ICTs are the most effective means of communication in disasters and in some situations the only means of communication with outside world due to the damage and destruction of traditional means of communication (Paradiso, 2010). ICTs enable filtering, processing and analysing huge volumes of data and help disaster managers in handling emergencies. ICTs facilitate communities to share information and knowledge and enable them to participate in disaster management activities along with other stakeholders.

Advancements in the areas of IT and E-Governance have trickled down to other areas including disaster management. It is evidenced from the variety of ICT applications created by various agencies related to disaster management in India. Yet, the opportunities provided by ICTs are yet to be harnessed. The main argument of the chapter is that the efforts made by the governments at the national and state level in harnessing ICT potential for disaster management are inadequate and lopsided. The chapter also argues that the applications of ICTs in disaster management are bound by certain prerequisites and challenges.

Advancements of ICT Applications in Disaster Management

The convergence of telecommunications and Information Technology (IT) and Information systems (IS) has resulted in what is called information and communication technology (Singh, 2013). Information and communication technology refers to numerous technologies and applications that are used to establish communication over a distance (telecommunications) and to store, process and retrieve data (information technology). ICTs cover any product that will store, retrieve, manipulate, transmit or receive information electronically in a digital form (Bowman et al., 2005). For example, electronic computers, radio, telephone; broadband, television, internet, robots etc. are all equipment which can be classified as ICTs. Information and communication technology play an important role in overcoming the limitations of time and distance to communicate, exchange information or work together.

ICTs affect the functioning of organizations fundamentally. ICTs help in restructuring and reforming organizations to make them more efficient. By replacing traditional ways of providing services and information with an electronic approach, ICTs can improve the quality of services within organizations. ICTs facilitate the integrated operation of organizational processes related to internal and external provision of information and services, decision making and management. ICTs also help in streamlining basic organizational data concerning all the stakeholders of an enterprise like managers, employees, customers etc. ICTs also facilitate integration of internal automation (back office) with communication (front office) between organizations and their environment. This integration is referred to as e-Business in general and e-Government in government organizations. E-Government claims to improve public service delivery, improve government's relations with citizens and enable citizen participation in governance (Bhatnagar, 2004; Prabhu, 2004).

Natural Calamities like excessive rainfall, floods, landslides, hurricanes, earthquakes or tsunamis become disasters when they damage and destroy life and property (Singh, Punia, Haran, & Singh, 2018). Disasters are the outcome of interplay between hazards and vulnerabilities (Medury, 2008). They are natural, technological and human initiated events that disrupt the normal functioning of the economy and society on a large scale (National Research Council, 2007). United Nations International Strategy for Disaster Reduction (UNISDR) defines disaster as a serious disruption of the functioning of a community or a society involving widespread human, material, economic or environmental losses and impacts, which exceeds the ability of the affected community or society to cope up using its own resources (UNISDR, 2009). Traditionally, disasters are categorized into natural and man-made. But of late, this categorization has become redundant as the frequency of occurrence of natural disasters is enhanced by human developmental activities. There is an intimate relationship between disasters and development. On one hand, unsustainable development practices increase the disaster risk and on the other, disasters stall development and disturb normalcy.

International Federation of Red Cross and Red Crescent Societies (IFRC) defines Disaster Management as the organization and management of resources and responsibilities for dealing with all humanitarian aspects of emergencies, in particular preparedness, response and recovery to lessen the impact of disasters. There are six phases in Disaster Management cycle. They are mitigation, risk reduction, prevention, preparedness, response and recovery (Wattegama, 2007, p. 5). Mitigation is an effort to reduce the impact of future disasters which involves designing, constructing, maintaining and renovating infrastructure to resist physical forces of disasters. It also involves efforts to decrease the exposure of life and property to hazardous conditions. Risk reduction is the set of actions and measures to avoid or reduce future risks due to disasters. Prevention is avoidance of disasters by taking certain measures. Preparedness means complete actions taken before the disaster to deal with anticipated problems of disaster management which includes training, capacity building, development and use of response and recovery plans and systems, public education, awareness and information. Disaster response covers all those efforts undertaken immediately once the disaster strikes. It includes dissemination of alerts and warnings; evacuation and other forms of protective action; mobilization of emergency personnel, volunteers and material resources; search, rescue and care of casualties and survivors, damage control and restoration of public services. Disaster recovery covers both short term and long-term activities aimed to restore the physical and social conditions to their pre-disaster state.

There is a paradigm shift in disaster management activities from post-disaster relief, response and recovery to a holistic management of disasters covering pre-disaster prevention, mitigation, preparedness, response, relief, recovery and reconstruction (Nair, 2012). Since complete prevention of disasters is not humanly possible, communities all over the world are trying to reduce the impact of disasters as much as possible by equipping themselves with proper information, resources and capacities in the name of disaster risk reduction. Disaster risk reduction is the concept and practice of reducing disaster risks through systematic efforts to analyse and manage the causal factors of disasters, including reduced exposure to hazards, lessened vulnerability of people and property, wise management of land and environment, and improved preparedness for adverse events (UNISDR, 2009).

Can India Manage Disasters without ICT?

India is susceptible to both natural and anthropogenic disasters due to her unique geo-climatic and socio-economic conditions. According to national disaster management division, 58.6% of land mass is prone to earthquakes, 40 million hectares of landmass (12% of landmass) is prone to floods, 8000 km of coastline is prone to cyclones and almost 68% of total geographical area is vulnerable to droughts and hilly areas are at the risk of landslides (MHA, 2011). The susceptibility of India to disasters increased manifold in the new millennium. India experienced World's deadliest disasters such as Odisha Super cyclone (1999), Gujarat Earthquake (2001), Indian Ocean Tsunami (2004), Bihar Floods (2007), Uttarakhand Flash Floods (2013), Kashmir Floods (2014) and Kerala Floods (2018) in the new millennium only. This necessitated India to put in place robust disaster management machinery and infrastructure for handling disasters.

National Disaster Management Division (NDMD) within the Ministry of Home Affairs is the nodal agency of disaster management in India. India used to follow the traditional disaster management model of post-disaster resilience. The focus was more on response, relief, recovery and rehabilitation rather than prevention, preparedness and mitigation. The administrative apparatus for disaster management was also designed accordingly as reflected in the nomenclature like relief departments, relief commissioners, relief codes/manuals etc. (MHA, 2011). This relief-based approach has given way to a more holistic prevention-based approach. The new approach is a result of the conviction that development cannot be sustainable unless disaster prevention and mitigation measures are integrated into the process of development. This approach has also been reflected in the National Disaster Management Act, 2005. The act provided a techno-legal framework related to institutional mechanisms, disaster prevention strategy, early warning systems, disaster mitigation, preparedness, response and human resource development. The act has also provided a plethora of institutions like National Disaster Management Authority, National Executive Committee, National Institute of Disaster Management, National Disaster Response Force, State Disaster Response Force, State Executive Committee and District Disaster Management Authority. The national policy on disaster management, 2009 also emphasized the prevention-based approach to disaster management. The departments of relief are renamed as departments of disaster management, and relief commissioners are renamed as 'office secretary, disaster management and relief' to institutionalize risk management in the system.

Information and Communication Technologies find a variety of applications in disaster management. ICTs like Internet, Geographical Information System, Remote Sensing and Satellite based communication system help a great deal in planning and implementation of disaster management. ICTs provide opportunities for large scale, automated and comprehensive collection of information about operations, decision-making and situational knowledge before, during and after a disaster. Such information can help the disaster managers to improve the understanding of the process of disaster management. It also helps in making available the experiences and lessons learned in real time so that decisions can be arrived at within less time. Information and Communication Technologies have provided capabilities that can help people grasp the dynamic realities of disasters more clearly and help them formulate better decisions more quickly (National Research Council, 2007). The adverse impact of disasters can be reduced by incorporating application of Information Technology tools and devices with disaster management policies (Kavitha & Saraswathi, 2018).

Information and Communication Technologies are used in almost all phases of disaster management (National Research Council, 2007). From disaster mitigation and preparedness to aftermath of a disaster, ICTs play vital role. Examples of ICT applications in disaster management are early warning systems (radio, television, telephone, SMS, broadcasting and

Internet), use of websites and blogs for disaster related information dissemination, use of data bases and call-centres to provide information about weather conditions, available material resources, personnel and volunteer organizations etc. and use of data mining tools to capture, analyse and share lessons learned from field experiences. Other technologies include Remote Sensing, Geographical Information System (GIS), Social Media, Big data, Cloud Computing, Robotics and Drones. Asian and Pacific Training Centre for Information and Communication Technology for Development (APCICT) in its module 9 titled 'ICT for Disaster Risk Management' has identified the following areas for the use of ICTs in disaster management.

Sl. No.	*Areas of ICTs for Disaster Management*
1	Collecting data and information in databases to manage logistics during emergencies as well as for mapping, modelling and forecasting.
2	Developing knowledge and decision support tools for early warning, mitigation and response planning.
3	Sharing information, promoting cooperation and providing channels for open dialogue and information exchange.
4	Communicating and disseminating information, particularly to remote at-risk communities.
5	Teaching, learning and raising awareness are all critical for developing a culture of disaster risk reduction, as well as building specific skill sets required by disaster managers.
6	Managing disaster risks by utilizing available ICT tools, including the internet, phones, television and radio, to alert communities of impending disasters coordinate response and rescue and manage mitigation programmes and projects.

Source: ICT for Disaster Risk Management (module 9), 2016, United Nations Asian and Pacific Training Centre for Information and Communication Technology for Development (APCICT), Incheon City, Republic of Korea

With the success of ICTs in the areas of governance, governments all over the world are trying to foster ICTs in disaster management activities. Still, the technological opportunities provided by ICTs have not been fully exploited by governments for managing disasters (NRC, 2007). There exists a digital divide between developed world and developing world in exploiting ICTs for disaster management. Whereas developed countries are advanced in harnessing ICT potential in improving community's capabilities to manage disasters, this is still in nascent stage in developing countries. While few governments have amended their disaster management policy frameworks to incorporate ICTs, most of the governments are yet to do so.

India is a leader among developing countries in designing, developing and adopting ICTs in governance (Bhatnagar, 2004; Pardhasaradhi, 2009; Prabhu, 2004). The advancements made in Information Technology and Telecommunications have been exploited by Government of India for developing suitable applications and systems for managing disasters. There are number of ICT based initiatives, tools and applications developed to help the disaster managers function efficiently. These applications and systems serve a variety of purposes in all phases of disaster management from prevention and preparedness to recovery and reconstruction. National Disaster Management Information and Communication System guidelines (2012) have highlighted ICT initiatives of Government of India vis-à-vis disaster management. The vision statement of these guidelines highlights the policy thrust of the national disaster management authority. The vision is to:

> Build disaster resilience in the society through creation of a state-of-the-art knowledge-based National Disaster Management Information and Communication System (NDMICS) to provide GIS based value-added information along with assured multi-services of audio, video and data to the various stakeholders for proactive and holistic management of disasters. Value added information along with data would be sent to the right people at the right time by establishing a reliable, dedicated and latest technology based, National Disaster Communication Network (NDCN), with particular emphasis on last-mile connectivity to the affected community during all phases of disaster continuum (NDM Guidelines, 2012, p. V).

The guidelines listed various ICT initiates for different disasters like ICT infrastructure, Hazard forecasting and early warning network, Emergency Operation Centre, national disaster management information system (NDMIS), The National Disaster Communication Network (NDCN), last-mile connectivity and other emerging trends.

Policy frameworks related to Disaster management such as National Disaster Management Policy, 2009 and National Telecom Policy, 2012 emphasized on the incorporation of ICTs. National Disaster Management Policy, 2009 highlighted the significance of ICTs in disaster management which is exemplified in the vision statement of the policy. The vision of the policy is 'to build a safe and disaster resilient India by developing a holistic, proactive, multi-disaster oriented and technology driven strategy through a culture of prevention, mitigation, preparedness and response'. One of the objectives of the policy is to establish technological frameworks to create an enabling regulatory environment and a compliance regime. The policy reiterated that implementation of disaster management strategy must include communication and sharing of reliable, fast and up-to-date geo-spatial information using IT infrastructure. It also mentioned that efforts should be made for setting up IT infrastructures consisting of required IT processes, architecture and skills. It also talked about establishment of National Emergency Communication Network for disseminating warnings and information to the affected communities.

A Spread of ICT Infrastructure in India

Several national and international agencies are working towards developing ICT applications for disaster management in India. Government of India and United Nations Development Programme (UNDP) jointly launched Disaster Risk Management (DRM) programme by using ICT tools for faster response, effective decision making and developing well informed practitioners. Indian Meteorological Department is the nodal agency for disaster forecasting and early warning of natural disasters in India. Besides IMD, Regional Specialized Meteorological Centre (RSMC) in New Delhi under the aegis of World Meteorological Organization monitors imminent tropical cyclones in Arabian Sea and Bay of Bengal with the help of India's geo-stationery satellite, INSAT. Early warnings are disseminated through a variety of communication channels such as radio, television, print media, electronic media, telephone, SMS, cell phone broadcasting, fax, telegram, police wireless network and internet.

Indian Space Research Organization plays crucial role in augmenting application of geo-informatics towards disaster risk reduction. The Disaster Management Support (DMS) programme developed by ISRO provides timely support and services from aero-space systems, both imaging and communications, towards efficient management of disasters in the country (MHA, 2011). Indian Institute of Remote Sensing (IIRS) and National Remote Sensing Centre (NRSC) of Indian Space Research Organization are mandated to develop technologies and applications to provide support for disaster management activities. NRSC has developed a web mapping service called 'Bhuvan' in 2009 that allows users to explore earth (especially India) in 2D/3D resolution. Besides, it provides disaster support services and free satellite data and products (NRSC, 2018). Indian Space Research Organisation (ISRO) also developed geo-informatics applications like National Agricultural Drought Assessment and Monitoring systems (NADAMS), INFRAS for forest fire monitoring, land mapping etc.

Ministry of Home Affairs in collaboration with United Nations Development Programme has developed an online inventory of emergency resources called India Disaster Resource Network (IDRN). India Disaster Resource Network is a web-enabled and Geographical Information System based national resource inventory for the collection and transmission of information about specific equipment, human expertise and critical supplies database at the district and state levels (IDRN, 2018). A GIS based National Database for Emergency Management (NDEM) has been initiated by ministry of Home Affairs in collaboration with departments/ministries such as department of Science and Technology and ministry of Communications and Information Technology. The database enables development of decision support system tools and establishment of ICT infrastructure that facilitate network connectivity, data ingest, validation, GIS databases organization, data dissemination and service hosting for disaster management (NRSC, 2018).

National Disaster Management Information and Communication System guidelines (2012) of National Disaster Management Authority, Government of India, have provided for a national disaster management information system (NDMIS) which is a geographically distributed network of users accessing the centralized national-level GIS repository for data and services (p. 46). NDMIS is based on the technical inputs like digital cartographic base, hazard maps related to various natural hazards and GIS database linked with GPS. NDMIS encompasses ICTs that are used for collection, storage, retrieval, mapping and analysis of geographic, demographic, topographic infrastructure details, socio-economic data for generating knowledge-based information and decision support systems. The guidelines have envisaged the development of NDMIS in a phased manner, with the advancement of technology, skills and expertise.

Ministry of Home Affairs has come up with a National Emergency Communication Plan to create National Emergency Communication Network employing both terrestrial and satellite-based communication technologies. The national emergency communication network at national, state and district levels uses POLNET (Police Communication Network) for multi-mode and multi-channel communication system. Similarly, National Institute of Disaster Management (NIDM) has created

India Disaster Knowledge Network (IDKN) as a part of South Asian Disaster Knowledge Network (SADKN). It is a web portal that offers array of resources and services such as knowledge collaboration, networking, maps emergency contact information system other valuable information related to disasters. It is envisaged as a network of networks created to cater to the needs of disaster managers, decision makers and other stakeholders (NIDM, 2018).

National Telecom Policy, 2012 recognized the significance of establishment of robust and resilient telecom networks for supporting disaster management activities by encouraging use of ICTs in prediction, monitoring and early warning of disasters and dissemination of information (NPT, 2012). It talked about Standard Operating Procedures, appropriate regulatory frameworks, priority routing of calls, single number based integrated emergency communication and response system. Accordingly, Ministry of Home Affairs has come up with an emergency number '112' in its National Emergency Response System (NERS) guidelines in the year 2015. National Emergency Response System (NERS) was initially started as an emergency response system to respond to women in distress (NERS guidelines, 2015). The guidelines have provided for extending the NERS to other emergency situations like natural disasters by integrating the emergency number '112' to other services like fire, medical, police, highway patrol etc.

Challenges

There are certain prerequisites, limitations and challenges in using Information and Communication Technologies for disaster management. The biggest technological challenge before the Government of India is to establish the national disaster management information system (NDMIS) and integrating it with data centres and Emergency Operation Centres at national, state and district levels for effective information management. Constant up-gradation of the hardware and software in tune with the emerging technologies and the maintenance of systems and databases require special attention and concern. Another significant challenge is the language barrier in the effective deployment of ICTs across all the regions of the country in disseminating information related to disasters. Development of content in local languages and their dissemination with the help of local communities would help in overcoming language barrier.

ICT, like any other instrument, can deliver its best when other necessary systems are in place. Successful adoption of ICTs in disaster management requires mix of political, cultural and institutional interventions, coordination between governments, market, civil society, media and other actors. Strong leadership, proper planning, stakeholder participation and a trained human resource pool, are conducive for effective use of ICTs in disaster management. ICT for disaster management is more about people and processes than about the technologies. The real challenge lies in identifying needs, gaps and capacities and assessing which technologies are required for achieving the targets set.

Another significant challenge is creating robust ICT infrastructure. ICT infrastructure (mobile networks, internet usage and access to mobiles, computers and other electronic gadgets) is critical for developing and adopting applications for effectively managing disasters. Better ICT infrastructure will act as foundation for governments to explore new possibilities and opportunities. However, huge investments are needed to create ICT infrastructure and the chances of immediate returns are meagre.

Human Resources and their capacity building are another crucial issue. Adequate emphasis should be placed on building human capacities to use these ICT tools and technologies. Professionals are to be developed in collecting information, processing and disseminating information and providing timely communication with affected communities. Lack of separate R&D facility to undertake innovative research for creation of better ICT applications for disaster management is another challenge. Limited budgets, lack of expertise and other resources, demographic differences (e.g., urban versus rural) and the press of routine responsibilities are major constraints for adopting ICTs in disaster management (National Research Council, 2007).

Conclusion

The opportunities provided by ICTs for effective management of disasters are vast. Emerging technologies like big data, cloud computing, mobile governance, satellite communications, fibre optics, internet of things (IOT), social media etc. have created immense possibilities for designing state-of-the-art applications for disaster management. Government of India is trying to harness the potential of ICTs by creating conducive policy framework, robust infrastructure and trained man power. The efforts made by successive governments since few decades have borne fruit which is evident from the fact that India could avert/lessen the adverse impacts of most of the disasters with the help of ICTs (NDM guidelines, 2012).

Although ICTs have immense potential, they are only means to achieve the goal of effective disaster management. The potential can be harnessed only when they are deployed with proper planning, under visionary leadership and capable workforce. Too much reliance on ICTs can also be counterproductive. There is every possibility of collapse of ICT infrastructure in times of disasters. During such untoward incidents, the damage is going to be more severe. Large disasters upset physical infrastructure, such as the electric grid, transportation and health care—as well as IT systems. IT infrastructures themselves need to be more resilient (NRC, 2007). ICTs cannot reduce risks and build resilience of communities. They only contribute to the improvement of disaster management activities. The real solution for the problems of disaster risk reduction and disaster management lies in building disaster resilient communities by equipping them with required resources.

References

Bhatnagar, S. C. (2004). *E-Government – From Vision to Implementation: A Practical Guide with Case Studies.* New Delhi, India: Sage.

Bouwman, H., van den Hooff, B., van de Wijngaert, L., & van Dijk, J. (2005). *Information and Communication Technology in Organizations: Adoption, Implementation, Use and Effects.* London: Sage.

Kavitha, T., & Saraswathi, S. (2018). Smart Technologies for Emergency Response and Disaster Management: New Sensing Technologies or/and Devices for Emergency Response and Disaster Management. In Z. Liu & K. Ota (Eds.), *Smart Technologies for Emergency Response and Disaster Management* (pp. 1–40). Hershey, PA: Information Science Reference (IGI Global).

Kuppuswamy, S. (2010). ICT Approaches in Disaster Management: Public Awareness, Education and Training, Community Resilience in India. In E. Asimakopoulou (Ed.), *Advanced ICTs for Disaster Management and Threat Detection: Collaborative and Distributed Frameworks.* Hershey, PA: Information Science Reference (IGI Global).

Medury, U. (2008). Toward Disaster Resilient Communities: A New Approach for South Asia and Africa. In J. Pinkowski (Ed.), *Disaster Management Handbook.* New York: CRC Press Taylor/Francis Group.

Ministry of Home Affairs, Government of India. (2011). Disaster Management in India.

Nair, S. S. 2012. *Geoinformatics Applications in Disaster Management*. Trainer's Module.

National Disaster Management Guidelines. (2012). National Disaster Management Information and Communication System (NDMICS), A publication of the National Disaster Management Authority, Government of India ISBN: 978-93-80440-12-5. New Delhi, India.

National Research Council. 2007. *Improving Disaster Management: The Role of IT in Mitigation, Preparedness, Response, and Recovery*. Available at: http://www.nap.edu/catalog/11824.html. ISBN: 0-309-66744-5. Accessed 21 Dec 2018.

Paradiso, M. (2010). Communications Models and Disasters: from Word of Mouth to ICTs, Italian Information Geographies, pp. 181–200. Available at: https://journals.openedition.org/netcom/373. Accessed 17 Jan 2019.

Prabhu, C. S. R. (2004). *E-Governance – Concepts and Case Studies*. New Delhi, India: PHI Learning Private Limited.

Singh, A. (2013). *A Critical Impulse to e-Governance in the Asia Pacific, With a Special Australian Case Study by Carol Johnson*. New Delhi, India: Springer Publications.

Singh, A., Punia, M., Haran, N. P., & Singh, T. B. (2018). Development and Disaster Management: A Study of the Northeastern States of India, Palgrave Macmillan. ISBN 978–981–10-8485-0 (eBook). Available at https://doi.org/10.1007/978-981-10-8485-0. Accessed 20 Dec 2018.

UNISDR. (2009). *UNISDR Terminology on Disaster Risk Reduction* (Geneva, United Nations). Available at: http://www.unisdr.org/we/inform/terminology. Accessed 12 Jan 2019.

Wattegama, C. (2007). *ICT for Disaster Management*, UNDP-APDIP, e-Primer for the Information Economy, Society and Polity. Available at: http://www.apdip.net/publications/iespprimers/eprimerdm.pdf

PART III

Building Community Resilience Through AI

CHAPTER 14

Can Community Plans Really Talk? Integrating and Strengthening Communications Through Artificial Intelligence

Thematic Division: AI in Local Governance

Andrew Estrain, Deepa Srinivasan, and Pat Pathade

Introduction

In recent years, Artificial Intelligence (AI) and Machine Learning (ML) have advanced to a state where they can be effectively utilized for many aspects and phases of disaster management. Many applications have already been successfully implemented in the areas of emergency response, and rescue, including during response and recovery efforts in the United States. For example, following Hurricane Harvey in Houston, Texas, autonomous drone technology allowed for high-resolution imagery for real-time mapping and damage assessments to take place at a rate that

A. Estrain (✉) • D. Srinivasan • P. Pathade
Vision Planning and Consulting, LLC, Fulton, MD, USA
e-mail: aestrain@vision-pc.net; dsrinivasan@vision-pc.net

T. V. V. Kumar, K. Sud (eds.), *AI and Robotics in Disaster Studies*, Disaster Research and Management Series on the Global South,
https://doi.org/10.1007/978-981-15-4291-6_14

would outpace ground-based rescue teams in locating survivors.[1] AI applications has also been used for prevention purposes, such as predicting future damages for a range of disaster scenarios, and in turn informing prevention methods and structures in need of reinforcement, also taking place in Houston, Texas.[2] As AI and ML have advanced to a state where they can be applied for many operations in disaster risk reduction, especially in response and prevention efforts, it is time to look at opportunities for AI and ML to be applied to the mitigation aspect of emergency management to serve people and government in mitigating risk and augment the applications that exist for prevention, response, and recovery.

Problem Statement

As natural disasters have increased in frequency throughout the U.S. and the world in recent years, there is an ever growing need to integrate hazard mitigation and risk reduction principles into the wide array of local community planning initiatives. Many planning documents exist within a jurisdictional planning framework, and there may be contradictions or gaps between various documents regarding hazard mitigation. There are several challenges, issues, obstacles, and inconsistencies to performing the plan review and integrating these plans manually, due to a high number of planning documents, inadequate manpower to perform the task, and varying plan update schedules. This paper will focus on using AI and ML for the integrated planning of hazard mitigation and risk reduction principles into existing jurisdictional planning frameworks and documents. Automating the plan review and integration process provides an opportunity to overcome the challenges, issues, obstacles, and lack of consistency resulting from performing the process manually.

Specifically, this paper aims to:

- Describe what hazard mitigation is and why it is needed;
- Highlight the importance, and provide examples of, integrating hazard mitigation (or risk reduction) principles into other local or community plans, and vice versa;
- Discourage planning 'silos' by identifying and emphasizing opportunities for interdepartmental coordination; and
- Explain efficiencies and benefits of automating the plan review and integration process through AI and ML.

What Is Hazard Mitigation or Why the Need for it?

Hazard mitigation is the effort to reduce loss of life and property by lessening the impact of disasters.[3] Different levels of governments engage in hazard mitigation planning to identify risks and vulnerabilities associated with natural disasters, and develop long-term strategies to reduce that risk, and to protect people and property from future hazard events.[4]

Hazard mitigation planning is a community-driven, six-step planning process that encourages jurisdictions to integrate mitigation with day-to-day decision-making regarding land use planning, floodplain management, site design, and other governmental functions and departmental policies. The hazard mitigation planning process is a phased approach that helps ensure compliance with federal regulations. Like most planning processes, hazard mitigation encourages and emphasizes overall intergovernmental and interdepartmental coordination, through the six-step planning process, depicted in Fig. 14.1.[5]

The Robert T. Stafford Disaster Relief and Emergency Assistance Act (1988), as amended by the Disaster Mitigation Act of 2000 (DMA 2000),

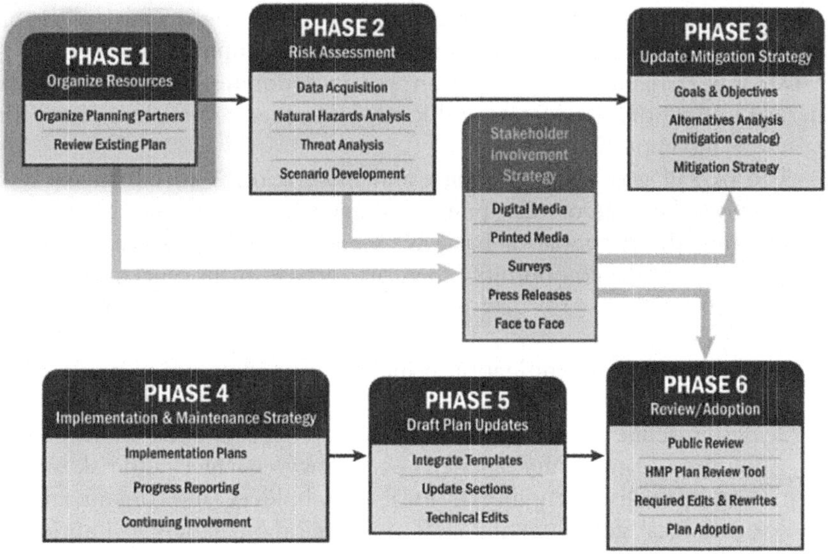

Fig. 14.1 Phases of the hazard mitigation planning process

is intended to 'reduce the loss of life and property, human suffering, economic disruption, and disaster assistance costs resulting from natural disasters.' Under this legislation, state, tribal, and local governments in the United States are mandated to develop a hazard mitigation plan to be eligible to receive various types of federal disaster funding and non-emergency disaster assistance through Hazard Mitigation Assistance Programs.

By developing a state, tribal, or local hazard mitigation plan, governments can identify specific hazard vulnerabilities, increase education and awareness regarding numerous natural and man-made hazards, and detail actions to mitigate those hazards. The hazard mitigation plan also lays out an implementation strategy which aims to properly focus resources on the greatest risks and vulnerabilities an area faces. It a mechanism to identify long-term risk reduction or hazard mitigation strategies, and a mechanism to build community partnerships, including various governmental departments, local organizations and businesses, and the public.

Plan Review and Integration

Plan review and integration is the process by which communities look critically at their existing planning framework and align efforts with the goal of building a safer, smarter, more resilient community. The process is a two-way exchange of information and incorporates concepts between hazard mitigation plans (state and local) and other community plans, for example, a comprehensive plan.

The overall goal of plan review and integration is to effectively link plans and policies across disciplines and agencies within a community by considering the potential of hazards as one of the key factors in future development of the community. There are three primary objectives to plan integration:

1. Integrate hazard mitigation principles into all community planning initiatives;
2. Better define roles of, and improve interdepartmental coordination between, community planners, emergency managers and risk reduction specialists, engineers, local stakeholders and organizations, other local governmental staff, and regional partners in improving disaster resiliency; and

Fig. 14.2 Hazard mitigation and local planning mechanisms

3. Ensure harmony between the hazard mitigation plan and other local planning mechanisms by identifying gaps and inconsistencies between policies, plans, and ordinances (Fig. 14.2).

There are two primary approaches to manually perform the plan review and integration. The first is to integrate hazard information and mitigation strategies, policies, and principles into already existing local planning mechanisms, and vice versa. This involves:

- Including natural hazard information such as past events, potential impacts, and vulnerabilities;
- Identifying hazard-prone areas throughout the community and reflecting that in other community planning initiatives;
- Identifying inconsistencies and gaps in existing local planning mechanisms; and
- Developing appropriate goals, objectives, policies, and projects to protect people and property and to reduce the overall impacts of natural hazard occurrences.

The second way to perform plan review and integration is by encouraging collaborative planning and interagency coordination. This involves collaborating across departments and agencies with key staff to help share knowledge and build relationships that are important to the successful implementation of hazard mitigation actions and programs.

Integrating Hazard Mitigation Principles into Other Local Planning Mechanisms (Integrate Through Plans)

Communities often have a large number of assorted plans, ordinances, and programs that shape their future and serve as road maps to development, sometime more than 15 documents. All relevant community plans and ordinances should be collected and analyzed to identify risk reduction principles through plan evaluation through a series of review questions within the community planning framework, where hazard mitigation can be integrated. Examples of questions for community-specific plans and ordinances, are presented below in Table 14.1:

[a] United States Department of Homeland Security, Federal Emergency Management Agency. July 2015. Plan Integration: Linking Local Planning Efforts. Washington, D.C.: United States Department of Homeland Security, Federal Emergency Management Agency. Retrieved from: https://www.fema.gov/media-library-data/1440522008134-ddb097c-c285bf741986b48fdcef31c6e/R3_Plan_Integration_0812_508.pdf

[b] United States Environmental Protection Agency, Smart Growth Implementation Assistance. July 2014. Flood Resilience Checklist. Washington D.C.: United States Environmental Protection Agency. Retrieved from https://www.epa.gov/sites/production/files/2014-07/documents/flood-resilience-checklist.pdf

[c] Ibid.

[d] United States Department of Homeland Security, Federal Emergency Management Agency. July 2015. Plan Integration: Linking Local Planning Efforts. Washington, D.C.: United States Department of Homeland Security, Federal Emergency Management Agency. Retrieved from: https://www.fema.gov/media-library-data/1440522008134-ddb097c-c285bf741986b48fdcef31c6e/R3_Plan_Integration_0812_508.pdf

[e] Ibid.

[f] Ibid.

[g] United States Environmental Protection Agency, Smart Growth Implementation Assistance. July 2014. Flood Resilience Checklist. Washington D.C.: United States Environmental Protection Agency. Retrieved from https://www.epa.gov/sites/production/files/2014-07/documents/flood-resilience-checklist.pdf

Table 14.1 Plan integration review questions

Land use
Does the land use plan include policies to restrict or regulate the density of new development in high-hazard areas or guide new development away from high-hazard areas?[a]
Has the community encouraged agricultural and other landowners to implement pre-disaster mitigation measures?[b]
Transportation and infrastructure
Does the plan identify critical facilities and infrastructure that are located in vulnerable areas and should be protected, repaired, or relocated (e.g., town facilities, bridges, roads, and wastewater facilities)?[c]
Does the transportation network provide redundancy (i.e. alternate routes) if certain key nodes or routes are affected by disaster?[d]
Emergency management
Does the community emergency operations plan show major evacuation routes that are prone to flooding or vulnerable to other natural hazards?[e]
Has the community adopted an evacuation and shelter plan to deal with emergencies from natural hazards?[f]
Environment/open space
Has the community adopted riparian and wetland buffer requirements?[g]
Does the plan encourage using green infrastructure techniques to help prevent flooding?[h]
Zoning ordinance
Does the ordinance prohibit development within, or filling of, wetlands, floodways, and floodplains?[i]
Do zoning or flood plain regulations require elevation of two or more feet above base flood elevation?[j]
Subdivision and land development ordinance
Do the subdivision regulations restrict the subdivision of land within or adjacent to identified natural hazard areas?[k]
Are there any specific mitigation measures included, such as additional setbacks in critical erosion zones, conservation of dunes or vegetation, floodproofing of facilities and utilities, and/or structural wind resistance and floodplain management?[l]
Building code
Are policies in place to reduce vulnerability to wind, water, hail, lightning, fire, and other pertinent hazards through regulating the location, size, design, type, construction methods, and materials used in structures?[m]
Are there measures for protecting vulnerable historically significant structures to preserve their historic character and appearance as well as protect them from damage from hazard events?[n]

[h] Ibid.

[i] Ibid.

[j] Ibid.

[k] United States Department of Homeland Security, Federal Emergency Management Agency. July 2015. Plan Integration: Linking Local Planning Efforts. Washington, D.C.: United States Department of Homeland Security, Federal Emergency Management Agency. Retrieved from: https://www.fema.gov/media-library-data/1440522008134-ddb097c-c285bf741986b48fdcef31c6e/R3_Plan_Integration_0812_508.pdf

[l] Ibid.

[m] Ibid.

[n] Ibid.

Examples of risk reduction strategies that may address and/or provide the answers to specific review questions include, but are not limited to:

- Incorporation of growth management techniques or promoting cluster development outside of high-hazard areas within the future land use plan;
- Creation of housing programs to retrofit publicly subsidized affordable housing to reduce damage after a disaster;
- Promotion of the restoration/protection of natural resources that buffer and help absorb floodwaters (wetlands); and
- Analysis of the current and future transportation infrastructure network to ensure adequate capacity during an evacuation.

These hazard mitigation and risk reduction-related principles are prime examples of mitigation and risk reduction strategies that, if aligned through all community plans, will enhance a community's overall resilience to hazards and disasters.

Coordinating with the Comprehensive Plan provides a mechanism to implement mitigation goals, objectives, policies, and actions included within the hazard mitigation plan.[6] In addition to the comprehensive plan, hazard mitigation and risk reduction principles should be integrated in other, more focused local plans and ordinances include: capital improvement plans, continuity of operations plans, emergency operations plans, transportation plans, stormwater management plans, building codes, floodplain ordinances, subdivision and land use ordinances, and zoning ordinances. Different community planning documents will likely have different goals, objectives, policies, actions, metrics, assumptions, etc., associated with the various planning elements. Ideally, a thorough plan review will result in a compilation of all items related to the specific planning discipline, and, in turn, generate a recommendation that brings all plans together.

Integration Across Agencies and Departments (integrate through people) Integrated planning also promotes interagency and interdepartmental coordination and brings to the table, representatives from different levels of government, and local government departments: development, police, budgeting, emergency management, environment, geographic information systems, forests and wildlife, health and family welfare, housing, information and publicity, irrigation and flood control, land and building, planning and urban development, public works, and transportation. Plan integration ensures collaboration and fosters interdisciplinary idea-sharing between agencies and departments,[7] as depicted in Fig. 14.3.

The process considers many diverse perspectives for reducing risk and promotes leveraging opportunities for partnerships to maximize resources and avoid duplication of efforts.[8] The integration process encourages collaborative planning and the breaking down of planning silos. It not only helps to break down the figurative walls between multiple disciplines and departments, but also works to build better multi-disciplinary partnerships around hazard mitigation.

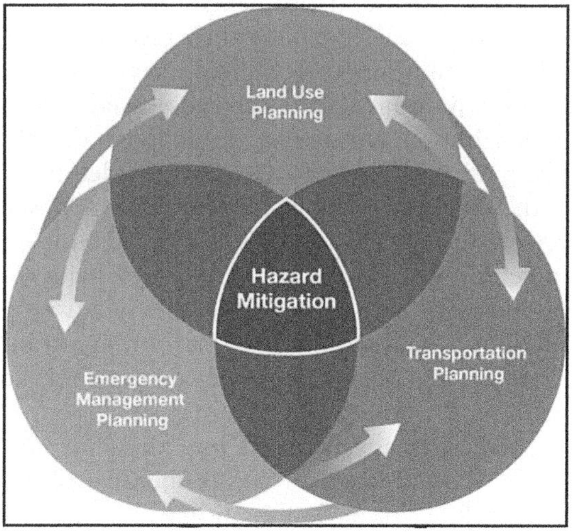

Fig. 14.3 Coordinating with local planning initiatives

The plan review and integration process provides the opportunity for people and departments to collaborate with one another, but that is generally easier said than done. Manually performing the plan review and integration process comes with several barriers, challenges, inconsistencies, and issues, all of which can be addressed and resolved through automating the plan review and integration process.

Challenges in Performing Manual Plan Review and Integration

While an integrated planning approach promotes overall interagency coordination and brings people and plans together to reduce a community's risk and to protect people and property from the impacts of natural hazards and disasters, there are still several challenges in this integration process and getting plans to 'talk' to one other.

There are often a large number of local planning documents, as there can sometimes be more than 15 separate plans and ordinances that are specific to an individual community, municipality, or jurisdiction. The majority of these planning documents, aside from the comprehensive plan, are specific to a certain discipline, and aligning each of these planning documents so they are in harmony with one another, is often a big challenge to performing plan review and integration, simply due to the volume of community planning documents.

These plans and ordinances are also often quite lengthy documents (300+ pages each) making their review, a time-consuming process. It may take several hours to complete a review of one plan. Performing a review of all community plans can take several days, or even weeks, placing an added burden on already-stretched staff. Those local governments and planning departments that have inadequate staffing resources to begin with, may not be able to effectively and successfully perform the plan review and integration process due to budget constraints. This may result in selecting only a small number of plans/ordinances for review, while excluding other pertinent documents. An automated plan review and integration process would ease that burden on local governments and would allow for each plan to be reviewed in a short amount of time, regardless of the length those documents may be.

With so many plans in use, they are inevitably updated on varying cycles. For example, a comprehensive plan may be updated every ten years, while a hazard mitigation plan is mandated to be updated every five years. These varying update cycles can create a lack of consistency between plan reviews, and with a large number of local plans and ordinances potentially in existence, they are often performed by the specific department responsible for developing or updating the document at hand. This often creates 'silos' within local government planning, meaning that operations are completed in-house (within one specific department), while reaching out to any other departments or individuals with expertise in specific topics. Automating the plan review and integration process would aim to break down those 'silos', allowing a variety of disciplines to work together as a team.

Another issue with manually performing plan review and integration is that there is no systematic way to complete the process; i.e., the methodology may vary from person to person, and from department to department. The lack of consistency between reviews based on the person performing the review, and the potential for omissions and human error, is one of the primary challenges during a manual review process. Automating the plan review and integration process would remove any inconsistencies between plan reviews and would eliminate the potential for human error.

Solution.

AI for Plan Integration

The introduction of AI and ML could effectively eliminate some of the major challenges, inconsistencies, and issues in performing the plan review and integration process. Automating this process will successfully and effectively enable community plans to talk and learn from one another, eliminate the potential for human error and lack of consistency between reviews, regardless of the total number and length of plans, or varying plan update cycles. Integrated planning can be performed through AI applications, in which a web application can be developed to perform tasks that normally require complex reasoning and/or human intelligence, and in this case, perform the integrated planning review of all relevant community planning documents.

Machine learning (ML) is an application of AI that provides systems the ability to automatically learn and improve from experience, without being explicitly programmed, and focuses on the development of computer programs that can access data and use it to learn.[9] The process of ML involves using observations to look for patterns in data and to make better decisions in the future.[10] The primary aim of ML for AI is to allow the computers to learn automatically without human intervention or assistance and to adjust future actions accordingly.[11]

This part of the paper explores how AI and ML can help communities, local governments, and associated agencies and departments look critically at their existing planning framework and align existing plans, goals, visions, policies, actions, etc., through an automated plan review and integration process. Performing integrated planning through AI requires a two-phase approach: Phase 1 – Build the Brain (BEB), and Phase 2 – Process the Plans (PEP).

Phase I: Build the Brain (BEB)

The first step is to 'Build the Brain' (BEB), which is comprised of preparing an Ontology, defining terms, and developing a knowledge base for plan integration and hazard mitigation. An Ontology deals with questions about what exists between different entities or data sources, or what can be said to exist, and how such entities can be grouped according to similarities and differences.

Prepare Ontology, Define Terms, and Develop a Knowledge Base

The first step is to prepare a common ontology for integrated planning from a hazard mitigation perspective, which includes defining domain-specific terms, or concepts. An ontology is a formal, explicit description of concepts in a domain of discourse, topics and sub-topics of each concept describing various features and attributes of the concept, and restrictions on those properties. An ontology together with a set of individual instances of concepts and topics constitutes a knowledge base (Fig. 14.4).[12]

In developing an ontology for plan integration from a hazard mitigation perspective, the various key concepts must first be defined. Examples of key concepts from a plan integration standpoint include, but are not limited to goals, objectives, policies, actions, metrics, assumptions, etc. Key concepts should then be arranged in a taxonomic hierarchy. These key concepts will be relevant to, and exist in some fashion, in all community planning documents. Once key concepts are identified, specific ontologies can also be defined. The ontologies, in this case, can be viewed as multiple disciplines within the community planning realm, such as housing, historic

Fig. 14.4 Plan integration related terms

preservation, land use, environment and natural resources, and/or transportation/infrastructure, while each ontology there will have associated goals, objectives, policies, actions, etc.

Once ontologies have been defined, different terms specific to each ontology can be defined using an Ontology Editor, such as Protégé, and the ontology stored in an application like Apache Jena. This involves identifying and defining domain-specific terms for the newly built ontologies. There are many terms within the planning discipline that are relevant and pertinent to plan integration for hazard mitigation. Hazard mitigation can be connected to all other community planning disciplines. It is recommended to develop a robust databank of terms associated with the various disciplines, that are also relevant to hazard mitigation and risk reduction. Table 14.2 below lays out examples of potential ontologies and associated key terms that may be found within specific planning documents, associated with those ontologies, and related to hazard mitigation and reducing risk. These key terms related to hazard mitigation are what a Community Planner would manually search for, throughout the various planning documents.

Table 14.2 Example ontologies and associated terms

Ontology Name	*Environment and Natural Resources*	*Land Use/Future Land Use*	*Transportation/ Infrastructure*	*Housing*	*Historic Preservation*
Key term examples	Streams, wetlands, floodplains	Conservation	Critical facilities	Building codes	Historic resources/ properties
	Watersheds	Zoning/rezoning	Evacuation routes	Elevation	Architectural features
	Steep slopes	High-hazard areas	Bridges and roads	Drainage	Cultural resources
	Stormwater management	Suitable/ unsuitable lands	Capital improvements	Building design/ construction/	Historic infrastructure
	Resource conservation	Density (development/ people)	Water resources	Infill development	Reconstruction/ rehabilitation
Ontology name	**Public facilities and services**		**Open space**	**Zoning/ subdivision and land development**	
Key term examples	Wastewater management		Protected lands	Buffering	
	Emergency operations (emergency services)		Sustainability	Erosion and sediment control	
	Public health and safety		Protective ecosystems	Sewage disposal	
	Sheltering		Conservation/ preservation	Impervious surfaces	
	Continuity of operations		Buffer requirements	Cluster development	
Hazard term examples	Flood, Fire, Wind, Tornado, Hurricane/Typhoon, Snow/Ice/Winter Storm, Hail, Landslide, Lightning, Earthquake, Drought, Extreme Heat				

Once specific terms are defined, an inbuilt knowledge base should be developed, based on years of plan integration and hazard mitigation experience. All captured knowledge may be viewed as a set of rules, as this data and knowledge base will be used to generate plan integration recommendations using the recommendations engine developed and utilized during Phase II.

Phase II: Process the Plans (PEP)

After building the brain, Phase 2 – Process the Plans (PEP), can be initiated. This phase involves natural language processing (NLP), indexing, and using a recommendation engine to generate plan integration recommendations.

Natural Language Processing (NLP)

NLP involves running the plan document through an NLP engine such as Google NLP, CoreNLP or Open NLP to recognize and extract Named Entities (NER). NER provides person names, organizations, locations, categories, and salience. Depending on the capacity limitations of the engine, the system may have to split the document into several smaller parts that are under the engine's capacity limitations, process each part, and then collate the results that meet a threshold salience value into a single response array for further processing.

Document Tagging Using Ontology

The results from NER are lemmatized, or sorted in order to determine the headword, under which other words are then listed,[13] and then searched across the custom Ontology for matching terms. The Ontology will be stored in Apache Jena and Protégé used to maintain updates to the Ontology. Any matches found are grouped by the Ontology Top Level Concept and used to tag the document. Thus, the plan document will be enriched with Ontology terms, and optionally converted into Resource Description Framework (RDF) format and stored in a triple store database.

RDF is a general method for describing data by defining relationships between data objects, and allows for effective data integration from multiple sources, like different community plans,[14] while triple store databases store data as a network of objects with materialized links between them, as

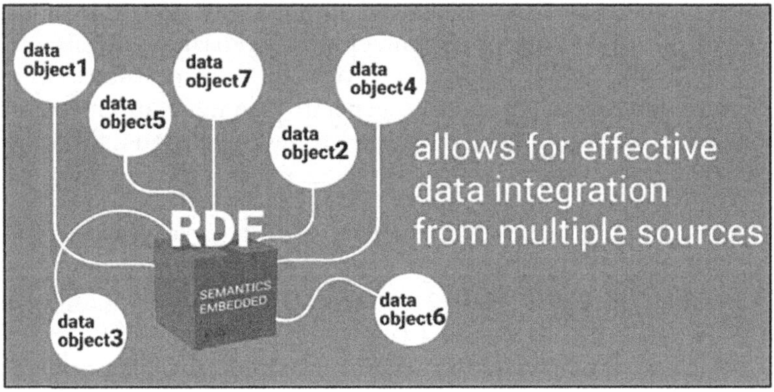

Fig. 14.5 RDF format

depicted in Fig. 14.5.[15] This makes RDF triple stores the preferred choice for managing highly interconnected data,[16] such as the wide array of assorted community plans and ordinances.

Recommendation Engine

A Recommendation Engine would be a custom, rules-based program where rules are generated using machine learning and training sets. Conceptually, the engine will be trained to find intersecting concepts and gaps across the assorted community documents that are enriched with terms from the Ontology. Using the intersection and gap knowledge, the Recommendation Engine would generate a summary report on how well the plans integrate, the gaps that need to be addressed, and recommendations on how to fill the gaps.

For example, the Recommendation Engine may indicate that a community's Comprehensive Plan is not well-integrated with the Hazard Mitigation Plan. The Recommendation Engine would go on to identify specific gaps between the two, or areas where the plans do not 'talk to each other', such as housing initiatives in the Comprehensive Plan contradicting floodplain management policies in the Hazard Mitigation Plan. After identifying specific gaps, the Recommendation Engine would generate a recommendation to align the two documents, such as recommending an edit for the Comprehensive Plan update to reflect up-to-date floodplain management policies called for in the Hazard Mitigation Plan.

Conclusion

Using AI and ML to automate the plan review and integration process will significantly address, with the goal of eliminating, the several major challenges, inconsistencies, and issues associated with performing plan review and integration manually.

Automating the plan review and integration process will result in capturing and retaining knowledge and would effectively ease the burden for local governments on completing this process manually. This automation, and capturing and retaining of knowledge, will save local governments and specific agencies/departments from spending large amounts of time and money (in the form of man hours), reduce the potential for human error, remove subjectivity and maintain consistency.

Automating this process will enable all community planning documents and initiatives to build off one another, regardless of the total number and length of plans, or varying plan update cycles. Using AI for plan review and integration will allow for specific gaps and inconsistencies within planning documents to be analyzed, identified, and addressed on a periodic or regular basis.

By following the two-phase approach of Building the Brain and Processing the Plans, it is possible for all planning documents to 'talk to each other', and effectively create an easier pathway for implementation of risk-related policies and actions. Automating the plan review and integration process through AI and ML will improve risk reduction and hazard mitigation, a phase in the emergency management cycle where AI applications have not been applied as frequently and successfully as during the response and recovery phases. Overall, automating the plan review and integration process through AI and ML will improve community resiliency.

While this paper focuses on planning efforts and initiatives within the United States, this approach to automating the plan integration and review process through AI and ML can be applied in India and other countries around the world. There are several levels of jurisdictions within the Indian Government structure (national, state, and local), each with numerous departments with specific responsibilities. This process can be applied to all pertinent documents within the planning framework to effectively enhance interdepartmental communication and coordination, while better preparing the jurisdiction for disasters by reducing risk and, in turn, mitigating direct impacts.

Notes

1. Diamandis, P.H. April 12, 2019. AI and Robotics Are Transforming Disaster Relief. Moffett Field, California: Singularity University. Retrieved from: https://singularityhub.com/2019/04/12/ai-and-robotics-are-transforming-disaster-relief/#sm.000bswipd14jacp2pak11h2ynyf24
2. Ibid.
3. United States Department of Homeland Security, Federal Emergency Management Agency. April 17, 2019. Hazard Mitigation Planning. Washington, D.C.: United States Department of Homeland Security. Retrieved from: https://www.fema.gov/hazard-mitigation-planning
4. Ibid.
5. No Author. 2016. The Hazard Mitigation Planning Process. Location Unknown: Mitigation Hazards. Retrieved from: http://mitigatehazards.com/planning-process/
6. Ibid.
7. Ibid.
8. Ibid.
9. Expert Systems. n.d. What is Machine Learning: A Definition. Rockville, MD: Expert System Enterprise. Retrieved from: https://www.expertsystem.com/machine-learning-definition/
10. Ibid.
11. Ibid.
12. Noy, N.F., McGuinness, D.L. n.d. Ontology Development 101: A Guide to Creating Your First Ontology. Stanford, California: Stanford University. Retrieved from: https://protege.stanford.edu/publications/ontology_development/ontology101-noy-mcguinness.html
13. Dictionary.com, LLC. n.d. Lemmatize. Unknown Location: Dictionary.com, LLC. Retrieved from: https://www.dictionary.com/browse/lemmatize
14. Ontotext. 2019a. What is RSF? Jersey City, New Jersey: Ontotext USA, Inc. Retrieved from: https://www.ontotext.com/knowledgehub/fundamentals/what-is-rdf/
15. Ontotext. 2019b. What is Triplestore? Jersey City, New Jersey: Ontotext USA, Inc. Retrieved from: https://www.ontotext.com/knowledgehub/fundamentals/what-is-rdf-triplestore/
16. Ibid.

References

Diamandis, P. H. (2019, April 12). *AI and Robotics Are Transforming Disaster Relief.* Moffett Field, CA: Singularity University. Retrieved from: https://singularityhub.com/2019/04/12/ai-and-robotics-are-transforming-disaster-relief/#sm.000bswipd14jacp2pak11h2ynyf24

Dictionary.com, LLC. (n.d.). Lemmatize. Unknown Location: Dictionary.com, LLC. https://www.dictionary.com/browse/lemmatize

Expert Systems. (n.d.). What is Machine Learning: A Definition. Rockville, MD: Expert System Enterprise. Retrieved from: https://www.expertsystem.com/machine-learning-definition/

Noy, N. F., & McGuinness, D. L. (n.d.). Ontology Development 101: A Guide to Creating Your First Ontology. Stanford, CA: Stanford University. Retrieved from: https://protege.stanford.edu/publications/ontology_development/ontology101-noy-mcguinness.html

Ontotext. (2019a). What is RSF? Jersey City, NJ: Ontotext USA, Inc. Retrieved from: https://www.ontotext.com/knowledgehub/fundamentals/what-is-rdf/

Ontotext. (2019b). What is Triplestore? Jersey City, NJ: Ontotext USA, Inc. Retrieved from: https://www.ontotext.com/knowledgehub/fundamentals/what-is-rdf-triplestore/

United States Department of Homeland Security, Federal Emergency Management Agency. (2015, July). *Plan Integration: Linking Local Planning Efforts.* Washington, DC: United States Department of Homeland Security, Federal Emergency Management Agency. Retrieved from: https://www.fema.gov/media-library-data/1440522008134-ddb097cc285bf741986b48fdcef31c6e/R3_Plan_Integration_0812_508.pdf

United States Department of Homeland Security, Federal Emergency Management Agency. (2019, April 17). Hazard Mitigation Planning. Washington, DC: United States Department of Homeland Security. Retrieved from: https://www.fema.gov/hazard-mitigation-planning

United States Environmental Protection Agency, Smart Growth Implementation Assistance. (2014, July). Flood Resilience Checklist. Washington, DC: United States Environmental Protection Agency. Retrieved from https://www.epa.gov/sites/production/files/2014-07/documents/flood-resilience-checklist.pdf

CHAPTER 15

The Challenge of Resilience in an Age of Artificial Intelligence

Acharya Shambhushivananda Avadhuta

The Coming Age of Artificial Intelligence

Science has already provided the possibility of liberation from hard labor. The digital revolution and AI are now promising to liberate us from the drudgery of repetitive chores. Artificial intelligence and machine learning capabilities are growing at an unprecedented rate. These technologies have many beneficial applications, ranging from machine translation to image analysis. Their relevance to strengthening disaster management capability is also being widely investigated (McKinsey Global Institute, 2018). McKinsey Global Institute recently reported the following:

> To coordinate and prioritize emergency response, governments and first responders must have an accurate and complete view of disaster zones. Frequent and broad area satellite imagery enables new AI-based systems to quickly and accurately detect infrastructure changes that may affect evacuation and response. AI can assist in improving relief efforts and emergency preparedness with greater accuracy and on a much larger scale than human

A. S. Avadhuta (✉)
Centre of Neohumanist Studies, Ydrefors, Sweden

T. V. V. Kumar, K. Sud (eds.), *AI and Robotics in Disaster Studies*, Disaster Research and Management Series on the Global South,
https://doi.org/10.1007/978-981-15-4291-6_15

> workers. Object detection software was applied to satellite imagery to detect flooded roads after Hurricane Harvey in 2017. Several AI capabilities, primarily in the categories of computer vision and natural language processing, are especially applicable to a wide range of societal challenges. These capabilities are good at recognizing patterns from the types of data they use, particularly unstructured data rich in information, such as images, video, and text, and they are particularly effective at completing classification and prediction tasks. (McKinsey Global Institute Discussion paper—December 2018, AI for Social Good)

Augmented Intelligence, now popularly called Artificial Intelligence, is here to stay. An Indian philosopher and sage, Shrii Shrii Anandamurtiiji expressed it very eloquently as early as 1959:'What the cosmic mind is doing today in a tangible manner will continue to be done by unit-minds on this earth in gradual steps'.

There are also, however, serious concerns about potential misuse of AI. A workshop conducted jointly by several institutions on 'the malicious use of AI' was held at Oxford in UK in 2017 and its report focused on identifying threats and ways of prevention and mitigation. 'Some concerns are directly related to the way algorithms and the data used to train them may introduce new biases or perpetuate and institutionalize existing social and procedural biases. For example, facial recognition models trained on a population of faces corresponding to the demographics of artificial intelligence developers may not reflect the broader population. Data privacy and use of personal information are also critical issues to address if AI is to realize its potential. Europe has led the way in this area with the General Data Protection Regulation (GDPR), which introduced more stringent consent requirements for data collection, gives users the right to be forgotten and the right to object, and strengthens supervision of organizations that gather, control, and process data, with significant fines for failures to comply. Cybersecurity and "deep fakes" that could manipulate election results or perpetrate large-scale fraud are also a concern'(Report, 2018). Realizing the gravity of potential misuse a center for data ethics and innovation has been established in UK as a non-governmental organization.

Technology is only as good as the people who employ it and their purposes. Recognizing the potential of harm that can be caused by any self-centered person, there is a call for AI Codes, AI Regulators and Ethical Framework to deal with issues of safety, fairness, transparency and collective good from the very outset. The questions being raised are:

1. Does the data reflect the population? Algorithmic bias, racial bias, gender bias, background bias etc.; lack of diversity in data sets; lack of diversity in modeling; lack of diversity of mind-sets—these are all matters of concern. Who should be the custodian of all data —private corporations or some public authority?
2. Even if data reflects the population accurately, does it mean that we should continue to treat it for perpetuity? What is fair?
3. Are there some things that AI and related technologies can do which we should never do? What is the ethical framework guiding humans and these technologies? Can incentives be built into AI that may lead policy makers or administrators to make wrong decisions?
4. Can AI tamper with the democratic institutions and ultimately erode the freedom of individuals in the society? What are the standards to which AI should be made accountable? What systems are in place to ensure the safety of critical data related to infrastructures and other sensitive domains?
5. Who should be made accountable for the 'invisible' complex networks of AI that would underlie the systems being employed by decision makers of different institutions? How can responsibility be attached to humans for the use of data or AI or related technologies?
6. In the task of rebuilding disasters areas, what is the role of 'culture'? In building smart cities—what about people? How AI can be used to preserve local history and local cultures?

This paper raises the following important questions:

1. Will AI become a successful tool of empowerment rather than exacerbation of disparities? What are some other simmering techno-ethical issues related to AI?
2. What is the best way to face the challenge of economic renaissance in the age of AI?
3. How can we move toward a resilient society and strengthen disaster management capabilities?

Disaster Preparedness and Resilience

Disasters or calamities, in a cosmological sense, are the result of the inherent propensity of macro-cosmic forces to bring back the eternal equipoise in all strata of existence. Neohumanism is a philosophical response to the need to incorporate the principles of balance in all domains of human existence. Disasters are usually ruthless, indiscriminately sapping the vitality of a society. Hence, disaster preparedness calls for establishing ground initiatives for enhancing resilience capability of each and every community. Communities need caring support systems as well as survival, and coping skills to handle emergencies resulting from disaster of any kind. Vulnerable groups like old people, children, flora and fauna and persons with disabilities are the greatest victims of a disaster, and require extra attention. Disaster preparedness or survival skills training should, however, begin from an early age. Hence, the need for linking disaster preparedness with youth programs and creating volunteer disaster relief teams led and empowered by youths in every village, city or town. Besides, can we build 'smart' infrastructure without 'smart' people? We also need to pay attention to improving capabilities of ordinary citizens to serve as 'wise and smart' people. Neohumanist Education is a response to this need. Furthermore, how can we preserve our local cultures while designing sustainable cities after disasters?

Besides the material challenge of dealing with the survival issues of disaster victims, there are long-term developmental aspirations that need fulfillment. The goal of disaster management capability, therefore, is to ensure quickest responses toward survival, growth and sustainability. In this task, technological assistance can go a long way to convert disasters into opportunities for resilient, sustainable growth. Development communities can only welcome AI as a tool for achieving the targets of millennium development goals (MDGs). Technologies are only as good as those who employ them. AI (Artificial or Augmented Intelligence) should be seen only as a support and not a substitute for NI (Natural or Native Intelligence).

Resilience is the stamina or ability of an organism to cope with and recover from shocks, calamities or disasters or unfavorable conditions; to flourish and grow despite all odds and limitations; and, to attain a state of collective flourishing. In development theory, resilience refers to the centripetal quality built into the policies, institutions and governance in order to counteract any structural vulnerability—local, regional, national or global. Resilience is intimately connected with both disaster preparedness as well as development strategy. One of the areas of disaster research is also to look for opportunities and ways to correct the wrongs or existing imbalances that compound the impact of disasters.

The French economist Thomas Piketty and Indian born USA Professor Ravi Batra have presented overwhelming evidence that accumulated wealth has typically grown at a rate that outpaces increases in workers' wages all over the world. The result, they argue, is an ever-widening gap between the extremely rich and everyone else. The use of advanced technologies lies at the core of these widening gaps (Sarkar, 1987; Piketty, Thomas, Capital in the 21st Century, 2014). Would AI further exacerbate these gaps? The solution may lie in first understanding the dynamics of biopsychology that breed such imbalances. The moral choice of adopting the path for the future will determine the end result.

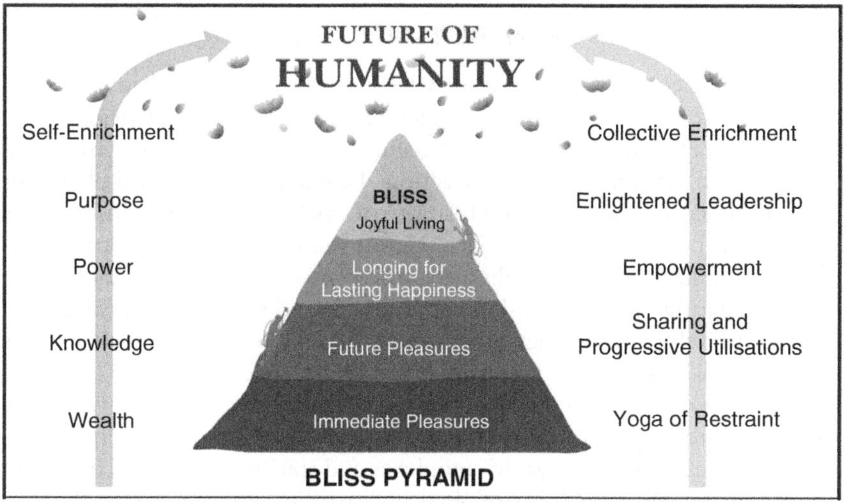

If AI can serve to reveal all disparities and help build a world of balance, of equipoise, of resilience, of sustainability, it would indeed be a day of celebration for all of humanity. On the other hand, if economic inequalities continue to persist and increase, AI could aid the process of translating these economic disparities into biological inequalities thereby threatening the very existence and worth of ordinary humans. Hence, ongoing research on technological impact assessment is needed and should be welcomed.

Prama: The Path of Resilience

Prama is a Sanskrit word meaning well-proportioned effort, a proper measure, balance, harmony, dynamic equilibrium and dynamic equipoise. The theory of Prama (Batra, 2003), as the guiding principle of a society, was first introduced by an Indian sage and philosopher Shri P.R. Sarkar in 1987 in order to strengthen resilience and to address the diverse problems of human society emanating from the inherent vulnerabilities in any system. A search for a proper path of progress in the midst of conflict-ridden situations is the crying need of our times. In Shrii Sarkar's view, 'the primordial natural state of dynamic equilibrium and equipoise of the Transcendental Entity is called *Prama-trikona* whose nucleus is *Purus'ottama*–Supreme Consciousness.' Disasters are reminders to us to correct all imbalances that have crept into different spheres of our life.

Loss of 'PRAMA' in Society

When the society loses prama (loosely translated as balance), chaos begins to set in. In the ultimate stage, the society ends up in the most degenerated doldrums stage. Society becomes directionless, without proper vision and without even a proper approach of action. This is the state of total blindness. The result could be total annihilation of civilization and culture. Mass destruction becomes commonplace. Nation-states reach the brink of bankruptcy. Conflicts increase, leading to widespread suffering of innocent citizens. Disparities abound and the purchasing power of common people declines to below poverty level. Such a state of affairs exposes the total lack of 'prama'. There is then dearth of structural balance (equipoise) as well as utter lack of dynamic equilibrium. At its climax, such situations give rise to mass discontent and even cause revolutions. Lack of prama in personal life brings physical diseases, psychic complexes and spiritual vacuum. The lack of prama in social life is the cause of physical disparities, dogma, and exploitative and imperialistic tendencies in the collective psyche, and spiritual bankruptcy.

Restoring PRAMA

In every sphere of life, there exists the expression of the threefold forces of nature: sentient (*sattva*), mutative (*rajah*) and static (*tamah*). The interplay of these forces is symbolized through the triangular representations called 'Trikonas'. Physical *prama-trikona* refers to the balance of physical expressions. The attainment of *prama* in the physical sphere is called *Prama-Samvriddhi.* Material prosperity of one and all is the natural outcome of *Prama-Samvriddhi.* Psychic prama-trikona refers to the balance in the psychic expressions. The attainment of prama in the psychic sphere is called *Prama-Riddhi.* The spiritual prama-trikona refers to the balance of forces in the spiritual stratum, its attainment being termed as *Prama-Siddhi.* The culminating resultant balance among physical, psychic and spiritual attainments is the eternal loka-trikona. The status of loka-trikona comes close to the primordial force of creation and is a pinnacled macro-cosmic stance of combined social-expressions. In loka trikona, the three sides of the cosmic triangle are represented by *prama-samvriddhi*, *prama-riddhi* and *prama-siddhi*. In his magnum opus *Ananda Sutram (*Anandamurtii, 1967*)*, Shrii Sarkar reveals the origins of endless guńa trikonas comprised of *sattva*, *rajah* and *tamah* attributes of the cosmic operative principle.

Sutra 4.1 'Triguńa átmiká sŕśti mátriká ashésa trikońa dhárá'.

[The tri-attributional primordial force (progenitrix of creation) flows on in endless triangular forms.]

All of creation is the result of three binding forces acting on the cosmic body of Supreme Consciousness in order to create endless mutual transformations through the process called homogenesis or homomorphic evolution. In the process of constant metamorphosis, a stage comes when due to the relative pressures of guńas, the balance of cosmic triangle of forces is lost and creation ensues through one of the vertices as *ichhabiija* or *kámabiija*. That starting point is termed as *Shambhuliunga* followed by *jinanashakti-náda* (vibrational flow without curvature) and *kalá-praváh* (with curvatures) in a sequential flow. The psychic and physical worlds are created until the creation reaches a nadir called *Svyambhuliunga*. The cosmic force lying quiescent around *Svyambhuliunga* is called *Kula-Kundalinii (jiivabháva)* and is the source of stamina and microcosmic power. The resilience of a group of people is intimately connected with the awakening of this dormant power lying within us. The endeavor to revert to a balanced prama-trikona and consequent resilience lies in building systems that are friendly to all beings- animate and inanimate (Avadhuta, 2017). Recognizing the role of microvita (smallest subatomic vital structures) and the self-controlling faculties in creation is the first step toward achieving equilibrium and equipoise in all strata of existence. Artificial Intelligence (AI), though a powerful tool, cannot be placed above the Natural Intelligence (NI) that is endowed to every living being.

The path of resilience based on prama is the shortest critical path to achieve better disasters-management capability as well as to attain the goal of sustainable development, thereby ensuring optimum utilization of resources while providing economic justice to the natural environment, future generations, vulnerable groups and all beings in general. This following section explores how we can bring about a resilient society based on this principle of seeking balance.

Survival (Asti): The First Step Toward Resilience

The three facets of resilience of a society are: Survival (*Asti*), Growth/ Development (*Bhati*) and Satisfaction (Anandam). These constitute the progressive goal that encourages forward- accelerated movement (*Dyut-Gati*). The three primary elements of pursuing the goal of ensuring minimum requirements with maximum amenities and consequent resilience in the mundane sphere, for instance, are

(i) Continuous Assessment of Community Needs
(ii) Commensurate Increase in Purchasing Power
(iii) Availability of Goods and Services

AI can assist us to respond more accurately in order to bring about equipoise of these three areas of physical-prama. Academic institutions could show leadership in showing the use of block chains, IoT, Cloud database and developing AI applications to create an accurate inventory of needs, availability of goods and services and commensurate purchasing power.

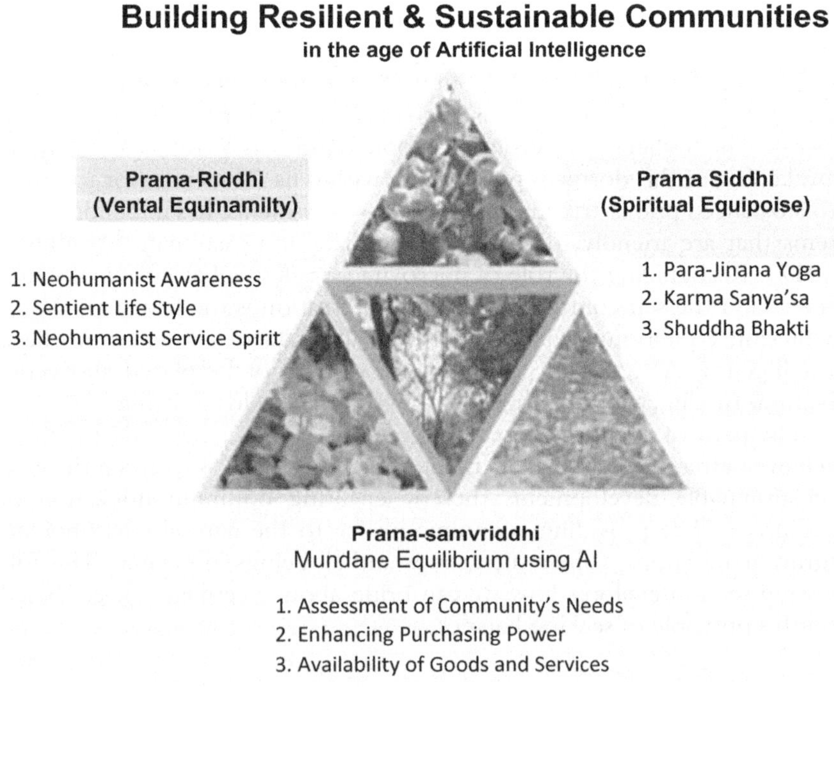

The task of a proper socio-economic system is to create conditions where there is maximum utilization of all resources, rational distribution of wealth, and where the utilizations are of progressive nature. Shrii P R Sarkar envisioned such a socio-economic system as early as 1959. In the words of Professor Sohail Inayatullah, 'The vision of **Pro**gressive **U**tilization **T**heory (**PROUT**) takes us beyond the dominant economic ideologies of capitalism and communism. Shrii Sarkar imagined a world with far greater efficiency, far greater productivity, far less inequality, living with nature and enhanced by amazing new technologies—"mind in technology", if you will—what we know today as the beginning of artificial intelligence (AI). This would be a planetary civilization where the boundaries would be functional not sentiment based as in today's nation-states. In this vision, contradictions do not magically disappear; however, exploitation decreases and the world gets better and better' (Inayatullah, 2019). It would encourage achieving regional self-reliance through a policy of economic decentralization, integrated farming as opposed to monocultures, use of renewable energy and green technologies, a healthy collaboration of public and private sectors, strong national security apparatus, stable and robust political leadership systems, integrated policy making, deliberate foresight planning, risk assessment and horizon scanning, clearly defined national goals, multiple streams of learning, while embracing psycho-social factors such as patriotism, justice, social identity, belonging and a sense of cosmic kinship.

Disasters can also be seen as opportunities provided by Nature to correct the lacunas of the past, remove the prevailing imbalances and achieve the goal of sustainability in the post- disaster phase. The developmental aspect of post-disaster management needs to be integrated with progressive developmental strategy of each area in order to ensure equitable growth and sustainability while removing the disparities and preserving the diversities. There is a need to constantly scout, locally and globally, for the most economical and sustainably progressive viable alternatives and make such options and information accessible for all needy parties.

The need for speed in post-disaster response also raises a set of issues of how best to utilize technological connectivity coupled with spirit of service in order to serve vulnerable groups especially children who are greatly traumatized after the disasters. Just as establishing fire brigades in a city is necessary to fight local fires, similarly support systems need to be built in every community to actively respond to all types of calamities. New communications technologies offer opportunities to create, strengthen and maintain real-world ties and enable rapid response to emergencies. Technology, if used benevolently, certainly has the power to strengthen the proximate community. Technology alone, however, cannot deliver a balanced and resilient society unless the human beings chose to act responsibly for collective good.

Advancements in technology must be balanced with enlightened leadership and an increase in compassion, magnanimity and wisdom among all citizens. Community-led disaster response centers linked with government support programs in every village, city or town could ensure that help would be readily available in any emergency situation as amply proved by AMURT (www.amurt.net) and similar NGO's.

The idea that resilience can be engineered and measured through technologies and methodologies in areas such as risk management, disaster management, forecasting, surveillance, biosciences, adaptive resource management and health sciences continues to dominate most approaches to resilience (Adaptation to Environmental Change, 2007). Some current examples include development of genetically engineered drought resistant crops; new water harvesting structures to overcome irregular rainfall and depleted water tables; improvements in national security and surveillance systems to thwart terrorist attacks, social unrest and crime; re-engineering of flood defenses, transport systems and buildings to withstand the effects of climate change; pre-assessment of health and other risks associated with adopting new technologies. It is now being increasingly recognized that psycho-socio-spiritual factors are equally important to successfully adapt and organize communities of people in a constantly changing and competitive global economic and ecological environment, for the sake of survival and prosperity. Any sustainable approach must be based on economic justice to the environment, needs of future generations, needs of all species, and especially all vulnerable groups. Advances in technology must be balanced with increases in compassion, magnanimity and wisdom among all citizens.

Allround Development (Bhati)

The development of a resilient society requires more than just mere application of technology for material ends. The moral and spiritual fiber of the nation also needs equal attention. While it may be easier to establish physical equipoise in the mundane sphere as some western nations claim to have achieved, it is more difficult to achieve mental equanimity. It is no surprise therefore that yoga and meditation have become household words in response to a compelling need to deal with the menace of 'stress' in the modern world. Thus, a proper socio-economic theory, neohumanist values, a spirit of service, cooperative mentality and survival skills contribute greatly to enhancing the resilience capability of a community. The endeavor to create a **GLOBAL NEOHUMANIST VISION** * can also inspire the younger generation to envision a world free from self-centered worldviews and myopic visions of the future.

> *** THE GLOBAL NEOHUMANIST VISION**
> [envisioning resilient sustainable local communities]
> where there is freedom, without fears; and,
> a constant endeavor for harmony among all species;
> where good health of all is the norm; and
> there is local sustenance: free from scarcities, poverty and disparities
> and, where purchasing power of all keeps improving;
> where conflicts are resolved through dialogues
> and challenges are faced with optimism and courage;
> where the uniqueness and diversities are celebrated;and, where ethics is the foundation of personal and social life;
> where science and technology are dedicated to greater welfare
> where higher-consciousness guides all forms of biological and AI;
> where religion and spirituality affirm cosmic kinship and rationality;
> where creativity, imagination, fine arts are for service and blessedness;
> and, where compassion, humor, joy and universal love pervade and reign!
> And, where deep-education (NHE)* inculcates and nurtures 'deeper understandings,
> cardinal values, innovations and leadership'; Relief Teams lend a ready helping hand
> in disasters; and The Renaissance Movement and PROUT** offer a new paradigm
> for self-sufficiency and economics of abundance and coordinated cooperation.
> *Neohumanist Education (NHE) ** Progressive Utilization Theory (PROUT)

Role of Intuitional Practices and the Goal of Anandam

Human beings came on this earth over two hundred thousand years ago and started a civilization about 10,000 years ago. We have still not successfully achieved a perfect balance in all spheres of human endeavor. This is due to a lack of psychic and spiritual prama-trikona. The presence of dogma and superstition has kept human intellect locked in senseless egoism. In the words of Dr. Marcus Bussey, 'Transformative harmony as an expression of *Prama* is not the end of tension but its benevolent expression. It is the pragmatic quest for balance between the two polarities of culture, one that is more inward looking and centralized and the other that is more resilient and receptive to learning. Harmony without dynamism is statis. Harmony as an element of social process is a normative goal that calls for social actors to reflect on their actions and their effects on the world around them.'(Culture, Harmony and Prama, 2013)

The benevolent luster of cosmic kinship and universal love will be able to express the grandeur of spiritual prama-trikona when there is a sweet blending of knowledge (*jiána*), action (*karma*) and devotion (*bhakti*). I would remiss if I failed to point out that traditional community practice of chanting kiirtans in India have long been believed to be a disaster-mitigating activity. Its importance should not be slighted, as declared by Shrii Caetanya Mahaprabhu and other spiritual masters.

> The physical afflictions of this material world are caused partly by nature, and partly by human beings themselves. Now, whatever might be the physical miseries—be they natural or man-made—if people collectively chant *kiirtana*, the calamities are dispelled. In case of natural calamities like flood, famine, drought or epidemic, or man-made calamities, miseries and tortures - if *kiirtana* is chanted with maximum sincerity, it will bring direct relief in no time. In addition, *kiirtana* removes the collective psychic afflictions as well - those which are already existing, and those which have not yet arrived but about which we have premonitions of their impending arrival. If *kiirtana* is done in advance, those impending troubles disappear. Why do they disappear? They are dispelled not merely because of the collective mental force of so many people, but also due to the impact of so many minds moving with tremendous speed under the inspiration of *Parama Purúsa*. At the place of *kiirtana*, not only the people who are themselves doing *kiirtana* will be benefited, but also those who are not participating—and even those who are not participating and who do not even like it—they will also be benefited! Those who listen sincerely to the chanting of *kiirtana* will be benefited and those who do not listen sincerely but simply hear inadvertently without any respect, they too will be benefited. (Shrii Anandamurtii, 1982, Kolkata)

Conclusion: A Call for Moral Advancement

In summary, the path of resilience is a constant endeavor to build balance in all strata of existence. Climate change resulting from solar activity, stratospheric ozone depletion, atmospheric aerosol loading, ocean acidification, biochemical flows, fresh water use, land-system change and biosphere integrity are some of the variables that need monitoring on a global scale in order to ensure maintenance of the Earth System (ES) in a resilient and accommodating state (Planetary boundaries: Guiding human development on a changing planet Persson, 2015). AI and Satellite Imagery together can help us become more alert or aware of the impending crises before they actually strike and should be maximally utilized to assess the vulnerable groups and risks involved in each case.

The ability to cope with any crisis also requires psychological/spiritual preparation and skills development. The responsibility to respond to the crises must also include the welfare of plants, animals and other life forms and even so-called inanimate things. For instance, it is being recognized that extensive use of pesticides has infiltrated into the underground water system in Punjab and could be the primary cause of cancer in the human population. Thus, the ongoing impact of technologies being employed and the assessment of risks to environment and life forms need to be continuously reassessed.

In the face of mass automation and artificial intelligence, the impending threat/promise is that we will all become productively superfluous. AI can never be a substitute for human intelligence, creative expressions, supporting relationships and wisdom. AI, some warn, could lead ordinary humans toward 'new addictions, apathy, indolence and boredom'. As we embark on using AI and related technologies in greater measure, we cannot lose sight of the equal need for moral development of citizens in every community and exemplary models of leadership at every level of society- from home to nations.

Technoethics (Gupta, 2016) as a living discipline must accompany all technological and scientific research. The issues of privacy, dignity, equity, diversity, disparities, health, safety, fairness, transparency, collective welfare, inner peace, moral advancement and elevation of human consciousness become increasingly important in the digital age of ICT (Information and Communication Technologies), AI (Artificial

Intelligence) and ML (Machine Learning). Age-appropriate course modules on 'The History of Moral Advancement' could thus be developed and introduced as a mandatory subject in curricula of all technical institutions. AI cannot be seen as a panacea without continuous efforts toward upgrading moral standards and elevating human consciousness.

Echoing the thoughts of Albert Einstein, 'we are concerned not merely with the technical problem of securing and maintaining outer peace. We also need inner peace and so we should also be concerned with the important tasks of education and enlightenment. When the ideas of humanity are war and conquest, these tools become dangerous. The fate of humanity is entirely dependent upon its moral development'(In Einstein's own words, n.d.). AI as a tool of mass surveillance and autonomous weapons has its scary side but as a tool to help actualize millennium development goals and build efficient disaster management capabilities, it is indeed a boon. Technologies become a curse or a boon depending on the degree of ethical responsibility employed in their use.

References

"Adaptation to Environmental Change: Contributions of a Resilience Framework" by Nelson, D. R. L., Neil Adger, V., & Brown, K. (2007). Annual Review of Environment and Resources, *32*, 395–419.

Anandamurtii, S. (1967). *Ananda Sutram.* Anandanagar, India: AM Publications.

Avadhuta, A. S. (2017). *Thoughts for a New Era.* Sweden: Gurukula Press.

Batra, R. (2003). *Common Sense Macroeconomics.* Dallas, TX: Liberty Press.

"Culture, Harmony and Prama" by Dr. Marcus Bussey, University of Sunshine Coast, *Gandhi Marg Quarterly* 35(1): 25–30 © 2013 Gandhi Peace Foundation, New Delhi. http://gandhipeacefoundation.org/

Gupta, Savita. (2016, November 18–19). Technoethics: From Society to Classrooms. In *Proceedings of International Educational Futures Conference*, eds. Dr. Shambhushivananda, & Dr. Sanjay Sharma, pp. 51–55. NITTTR-NERI Joint Initiative, Chandigarh, India.

In Einstein's own words. (n.d.). https://www.youtube.com/watch?v=_TmlYGdBodQ

Inayatullah, S. (2019). (UNESCO Chair in Future Studies) p. xi & Dr. Shambhushivananda p. 203–220. In Mulay, A. (ed.) *Economic Renaissance in the Age of Artificial Intelligence* (p. 313). New York: Business Expert Press.

McKinsey Global Institute. 2018, December). *AI for Social Good*–Discussion Paper.

Piketty, Thomas. (2014, April). *Capital in the 21st Century.* Harvard University Press, USA *Rising Inequality and Globalization*, 2017. https://www.youtube.com/watch?v=YtIw-n7z3VY

"Planetary boundaries: Guiding human development on a changing planet Persson." (2015, January 15). Veerabhadran Ramanathan, Belinda Reyers and Sverker Sörlin Stephen R. Carpenter, Wim de Vries, Cynthia A. de Wit, Carl Folke, Dieter Gerten, Jens Heinke, Georgina M. Mace, Linn M. Will Steffen, Katherine Richardson, Johan Rockström, Sarah E. Cornell, Ingo Fetzer, Elena M. Bennett, Reinette Biggs, https://doi.org/10.1126/Science.1259855 originally published online *Science* 347 (6223), 1259855.

Report: *The Malicious Use of AI: Forecasting, Prevention and Mitigation*. (2018, February). Oxford, UK.

Sarkar, P. R. (1987). *Prama*. Anandanagar, India: AM Publications.

CHAPTER 16

AI in an Urban Village in Delhi

Natasha Goyal and Anurag Singh

Introduction

The 2010 report by World Bank and United Nations titled 'Natural Hazards, Unnatural Disasters' highlight the social construction of the devastation which disasters expose. The frailties are the result of cumulative effect of individual and community decisions on issues such as land management, negligence of construction laws and regulations and lack of social integration. Lack of preventive action plans, resources and community resilience contribute to prolonged and delayed adverse effects on the environment and increase social vulnerability. The risk gets aggravated when the capacity of the community, in terms of social and economic infrastructure is compromised vis a vis the vulnerability and the hazard present in the region.

Priority 2 of the Sendai Framework for Disaster Risk Reduction (2015–2030) emphasizes strengthening Disaster Risk Governance

N. Goyal (✉)
Centre for the Study of Law, Governance & Disaster Research, Jawaharlal Nehru University, Delhi, India

A. Singh
Birla Institute of Technology and Science, Pilani, Pilani, RJ, India

T. V. V. Kumar, K. Sud (eds.), *AI and Robotics in Disaster Studies*, Disaster Research and Management Series on the Global South,
https://doi.org/10.1007/978-981-15-4291-6_16

through investment in structural and non-structural Disaster Risk Reduction (DRR) measures. Rapid strides in modern technologies like Artificial Intelligence (AI) and Machine Learning (ML) have made it feasible for their widespread application in disaster management cycle- from prediction, identification and classification to mitigation. These data driven modern technologies can be leveraged for vulnerability mapping of critical infrastructure and timely assessment of vulnerability of physical infrastructure, build robust early warning and response centric communication systems. Community engagement and participation can further prove crucial to make these data-driven technologies more robust, need based and real time facilitating efficient and transparent decision making and capacity building to deal effectively with the disaster. This paper would discuss the applicability and effectiveness of using Artificial Intelligence in mapping of risks in Munirka village using local participation for increasing institutional preparedness and building community resilience.

Vulnerability of Urban Villages: Case Study of Munirka

Delhi is located in Seismic Zone IV, which is classified as High Damage Risk Zone under the Earthquake Vulnerability Atlas of India (Vulnerability Atlas of India, 2019). The city is densely populated and has numerous informal and unauthorized settlements. Munirka is one such area, which is an urban village located in South West Delhi. Initially classified as a Lal Dora (*abadi deh*) area, the region is characterized by mixed land use. Since the area was exempted from the jurisdiction of building bye-laws under the 1963 notification by the Municipal Corporation of Delhi, it has witnessed unregulated growth of buildings and occupants. Unauthorized commercial activities (in violation of the Master Plan) in residential premises are an evidence of all possible violations of building bye laws and safety standards. Since most of the inhabitants belong to Low and Middle Income Group category, these commercial activities are undertaken in residential spaces to supplement the household income. Violation of Floor Space Index (FSI) norms can be witnessed through mushrooming of indiscriminate constructions (in both horizontal and vertical manner) amidst constricted congested lanes and buildings. Demolition of such structures becomes difficult due to prevalent housing scarcity and political pressures in a democratic political set-up (Fig. 16.1).

Fig. 16.1 (a) Commercial Establishments found in residential premises in Munirka. Such mixed land use is common in Lal Dora areas. (b) High Vulnerability due to low hanging high voltage electric wires and five-story vertical building constructions in Munirka village. (Source: Author's)

Further, this unregulated and haphazard manner of urbanization, without prior mapping of infrastructure and carrying capacity, has resulted in increasing vulnerability of these areas toward multiple hazards. The lanes are extremely congested due to the housing and commercial activities, making it unviable for any evacuation vehicle to respond timely due to lack of access. High voltage electric cables hang low, making the settlements prone to electrocution, especially during the monsoon season. Encroachments over the drainage system have compromised the drainage capacity, leading to water logging and stagnation. The area is highly prone to fire hazards, earthquake and flooding, with exposure getting accentuated due to densely populated settlements and lack of social infrastructure such as public hospitals, government schools etc. There are many illegal colonies in which law and order becomes a major challenge in a disaster scenario, due to high risk of human trafficking. Building collapse is usually witnessed due to the inability of weak structures to bear the pressure of additional floors constructed vertically, in violation of safety norms and codes. In a scenario where such events happen without any tremors due to geological reasons (Fig. 16.2), it becomes imperative to assess the vulnerability of people and infrastructure, before any hazard manifests as a large-scale disaster. Since demolition of physical infrastructure is unviable in a country inhabiting residents with low per-capita income, vulnerability mapping can be a pivotal step in assessment of the preparedness and capacity building. The following section discusses the potential of technology interventions such as aerospace imaging, vulnerability mapping through crowd-sourcing and Machine Learning, for effective risk mitigation and capacity building.

Fig. 16.2 Building collapse in Munirka Village on 23 February 2020 even in absence of tectonic activity underscores high vulnerability of physical infrastructure due to high rise building constructions on weak foundation. (Source: Author's)

Vulnerability Mapping through Aerospace Imaging

For generating the vulnerability map for an informal settlement, aerospace imaging puts emphasis on durable housing as an indicator. The parameters such as type of house- size, roof material and quality of construction material used, orientation of houses and physical infrastructure like road networks, area under vegetation etc. are taken into account. The hazardous locations, vulnerable to natural hazards are identified using domain knowledge and data on parameters such as Digital Elevation Models, rainfall, slope stability, nature of soil, drainage systems etc. The data for these geospatial technologies can be accessed through internet services like OpenStreet Maps, Google Earth/Google Maps, ISRO's Bhuvan server etc. Open source software packages (such as QGIS, ILWIS etc.) and advanced methods for data fusion using algorithms like dense image matching etc. can also be used for better imaging (Table 16.1).

There are various image characteristics which are crucial in the dataset which is collected. The model needs images with high spatial resolution (preferably < 1 meter). It means that the image pixel represents a smaller area on the ground for identification of smaller objects in the images. Medium to low temporal resolution is the second feature, to factor in the frequent change of demographics and infrastructure in an area, thus making the data more real time and authentic. The third characteristic is high spectral resolution (using sensors which can define an area over fine wavelength intervals) to effectively map the different elements in topography like vegetation, open spaces, buildings, clutters, etc. Such dataset forms the first stage in the process of vulnerability mapping (Fig. 16.3).

Table 16.1 Differences between physical infrastructure of unplanned/informal areas and planned areas which the algorithm takes into consideration

Characteristics	*Planned areas*	*Unplanned/Informal areas*
Size of buildings	Larger building sizes	Smaller (substandard) building sizes
Density	Low- moderate density areas, Less congestion Open spaces within or in the vicinity.	High density, high congestion Lack of open spaces
Layout	Layout pattern often in compliance with a master plan	Organic layout structure, non-compliance with planning standards

Fig. 16.3 Satellite view of Munirka Village. (Source: Google Maps)

Data-driven AI technologies also effectively substitute for large scale field surveying which is time consuming, expensive and may lack data accuracy. Satellite imagery can be leveraged to create digital elevation models, which can help to identify buildings which are prone to high risks. This imagery can be coupled with drone imagery to identify the rooftop material and to assess the underlying construction material and techniques, which are relatively more vulnerable to disasters such as seismic activity. High resolution images from satellites and drones can further be overlapped with street-view images from 360 degree street cameras to produce maps of physical structures based on their risk susceptibility. Such geo-spatial GIS mapping has been done for Mumbai city, by the Disaster Management Unit, Municipal Corporation of Greater Mumbai. Various critical infrastructures such as fire stations, police stations, hospitals, educational institutions have been mapped using satellite images and street-view images from more than 5000 CCTV cameras installed across the city. Important information such as responsive capacity (in terms of trained human resource, rescue equipment etc.), emergency contact numbers of nodal officers have been collected and fed into a centralized information management system using GIS mapping. This data repository becomes crucial for an efficient command and control system in case of any disaster. For example, in case of an incident of fire in an establishment, all the stakeholders (the nearest fire station, police station, hospital) in the buffer zone of 300 meter are alerted from the Emergency Operation Centre at Disaster Management Unit, MCGM for efficient and coordinated response. In case of an incident of more intensity, stakeholders in the buffer zone of 500 meter are sent alerts for timely action. As the next step, the city administration is mapping geological and structural vulnerability to hazards such as landslides, flooding and earthquake in different localities across the city for better risk assessment and preparedness.

Crowd-Sourcing Data from Communities: A Participatory Approach

Norris, Stevens, Pfefferbaum, Wyche, and Pfefferbaum (2008) define community resilience as a process that links network of adaptive capacities, viz. Information and Communication Technology (ICT), social capital, economic resources and community competence. This process requires engagement of people in risk mitigation, creation of organizational linkages for effective organizational behavior, efficient decision making skills, and robust channels of information production and dissemination which can continue to function in the face of disruption of normal activities. It is hence an amalgamation of community's knowledge, capacities and skills (Coles & Buckle, 2004). If the adaptation fails due to incapacity, resulting in persistent dysfunction, it is termed 'vulnerability'. If adaptation to altered environment occurs, it is called resilience. The presence of inter-organizational networks in the community, characterized by supportive community interactions and cooperative decision-making processes forms a crucial aspect (Goodman et al., 1998: 260). Participation of citizens, which helps to overcome the deficit of transparency and accountability, forms the pivotal and fundamental principle for community resilience.

In order to make mapping technologies more efficient and responsive to community needs, the data needs to be upgraded through crowd-sourcing, which involves different stakeholders such as local government, residents, students and academic institutions. The project can be implemented through Urban Local bodies and Land and Revenue Department, after adequate training in operating and handling GPS devices and mobile data collection. People from knowledge centers and educational institutions can be incorporated, who can further train local people to participate effectively in mapping. Tools such as online maps, atlas, mobile applications and GPS handheld units would play a key role in assisting these stakeholders to utilize the already available risk information in the public domain while collecting data on amenities and households.

Crowd-sourcing data also helps to add vital details to vulnerability maps, which may have been omitted in satellite imaging. Mapping of major critical and important infrastructure points such as shelter homes, hospitals, schools, fire stations, police stations, administrative buildings etc. can be done with more precision. Greater granularity of data can help in more effective and targeted relief effort in case of a disaster. It also helps to understand the community nuances like what is needed, when and why, which gets unaccounted for in technology. Such household survey by digitization volunteers also helps to collect data on access to public services and build strong data analytics capability in the targeted area. This dataset can be synced with the existing layers in geo-spatial data collected through satellites, to facilitate authenticity and transparency of data.

Implementation of this evidence driven urban resilience intervention can also prove crucial to map economic indicators such as legal status of property titles, source of income, penetration of financial insurance etc. It can also be leveraged to register basic building types, assess the transport pattern, collect basic demographic information and extensively document people's experiences of recurrent disasters such as fire, localized flooding etc. The training of people in operating GPS enabled digital devices and mobile data collection can facilitate the creation of data communities. Such an exercise can help to increment and digitize the pre-existing data on one hand, while simultaneously building community resilience through participatory vulnerability mapping. It can also help to enhance social capital across narrow identities of caste, religion and ethnicity.

Generating Classification Maps: Deep Learning Approach Using Support Vector Machines (SVM) and Feature Extraction

In the next stage, the data collected through geo-spatial imaging and crowd-sourcing is used to train a Machine Learning algorithm. For data processing, we use a classification algorithm called Support Vector Machine (SVM). SVMs use kernel functions (a class of algorithms that are used for analysis of patterns) to explain the nonlinear differences that exist between training samples. SVMs are suitable in this situation which has limited reference points, can handle large numbers of input features and learn

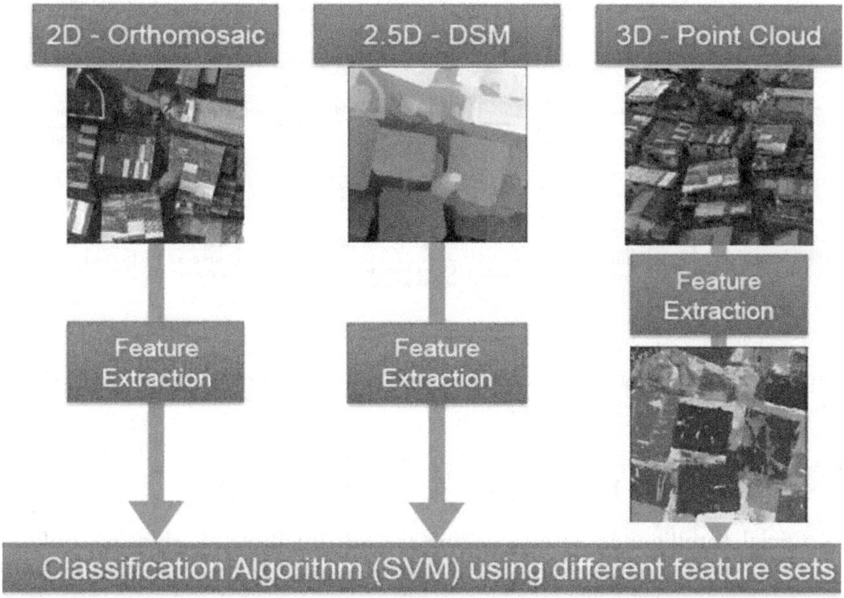

Fig. 16.4 Extracting spatial-contextual features from the geo-spatial image using deep learning. (Source: Bergado, Persello, & Gevaert, 2016)

nonlinear relations between these features. Feature extraction is an important engineering process as it provides the main information about the raw data identified by the algorithm, which is aided by high quality geo-spatial and crowd-sourced data (Figs. 16.4 and 16.5).

Thus, a hybrid approach involving community and modern data-driven AI based technologies, helps to understand the priorities of different disaster zones in a targeted manner so as to enhance the efficiency and effectiveness of the entire disaster management cycle. The perception of disaster risk varies in a community depending on social class, education, age, gender, etc., and thus risk assessment and risk reduction planning processes should be calibrated in accordance with the different perceptions, to ensure that the process becomes more inclusive. This process also helps to augment the social capital and community responsiveness to respond effectively as first time responders and be resilient in a scenario of an unpredictable event.

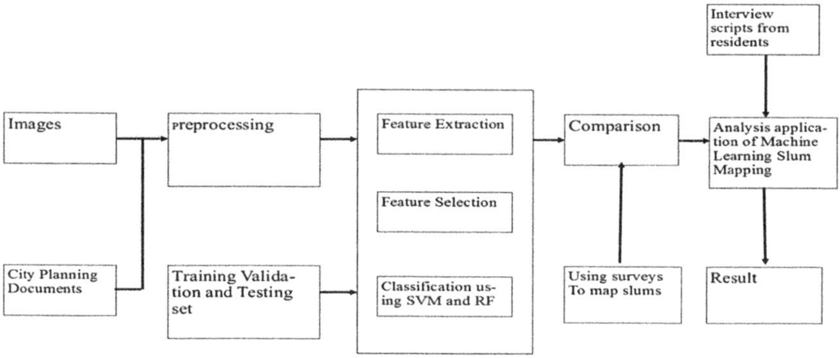

Fig. 16.5 Collating aerial mapping and crowd-sourcing data to generate vulnerability maps. (Source: Author's)

Limitations in the Use of AI

Artificial Intelligence based systems are not infallible, as they are only as smart as the data served. Availability of quality and authentic data-base becomes crucial for effective usage. Data storage and safety from cyber-attacks and data leakage becomes critical in the age of Information and Communication Technology, as such data can be mis-used from the aspect of national security. Installation of local servers and communication systems hence becomes very important for holistic and robust disaster preparedness. Proper guidelines and frameworks need to be put in place before undertaking any such exercise. Training of human resource becomes another imperative in a skill deficit country such as India, where scarce resources need to be diverted for various aspects such as health, education, poverty alleviation, rural agricultural distress, food security etc. as a priority. Writing robust algorithms and models free from algorithmic biases is another crucial challenge in using the AI based systems. The system has to be ready to adapt to new information during its lifecycle and change direction in accordance with real time environments.

Conclusion

South Asia is characterized by several 'urban village' type settlements. Munirka is one of the several such unplanned settlements, with dense and haphazard urbanization pattern. Maximization of private spaces at the cost of public spaces has exposed the infrastructure to natural as well as man-made hazards. Data collection using modern technologies and generating actionable intelligence through AI/ML based applications can be leveraged to assess the vulnerability of physical infrastructure and prior assessment of critical infrastructure which can provide necessary services such as shelter, medical aid and educational services in case of a disaster. These interventions can help to provide the required impetus for government agencies to fine-tune their planning and strategies for better response. Universality and deep penetration of technology such as mobile and internet can help to overcome the constraint of inadequate human resource in government institutions. Artificial Intelligence can also help in deducing patterns from data to identify vulnerable areas and populations and help model future requirements based on variables like population growth, rate of infrastructure development, changes in demographic patterns etc. Technology is by no means an end in itself. Hence, collaborative aspect becomes crucial for building effective community resilience.

References

Bergado, J. R., Persello, C., & Gevaert, C. (2016). A Deep Learning Approach to the Classification of Sub-Decimetre Resolution Aerial Images. In *Proceedings of International Geoscience and Remote Sensing Symposium (IGARSS): Advancing the Understanding of Our Living Planet* (pp. 1516–1519).

Chattopadhay, S., Dey, P., & Michael, J. (2014). Dynamics and Growth Dichotomy of Urban Villages: Case Study Delhi. *International Journal for Housing Science, 38*(2), 81–94.

Coles, E., & Buckle, P. (2004). Developing Community Resilience as a Foundation for Effective Disaster Recovery. *The Australian Journal of Emergency Management, 19*, 6–15.

Expert Committee. (2007). *Report of the Expert Committee on Lal Dora and Extended Lal Dora in Delhi*. Accessed from http://delhi-masterplan.com/wp-content/uploads/2009/09/Expert-comittee-report-on-lal-dora.pdf

Goodman, R., Speers, M., McLeroy, K., Fawcett, S., Kegler, M., Parker, E., et al. (1998). Identifying and Defining the Dimensions of Community Capacity to Provide a Basis for Measurement. *Health Education & Behavior, 25*, 258–278.

Norris, F. H., Stevens, S. P., Pfefferbaum, B., Wyche, K. F., & Pfefferbaum, R. L. (2008). Community Resilience as a Metaphor, Theory, Set of Capacities, and Strategy for Disaster Readiness. *American Journal of Community Psychology, 41*, 127–150.

Sendai Framework for Disaster Risk Reduction 2015–2030. (2015). UNISDR. Accessed from http://www.wcdrr.org/uploads/Sendai_Framework_for_Disaster_Risk_Reduction_2015-2030.pdf

Singh, A. (2018). Critical Pedagogy of Disaster Research. *Dialogue, 19*(3), 33–46.

The Disaster Management Act. (2005). *National Institute of Disaster Management.* Delhi: Government of India. Accessed from https://ndma.gov.in/images/ndma-pdf/DM_act2005.pdf

Vulnerability Atlas of India. (2019). *Building Materials & Technology Promotion Council (BMTPC)*, Ministry of Housing and Urban Affairs, Government of India.

World Bank. (2010). *Natural Hazards, Unnatural Disasters. The Economics of Effective Prevention.* Washington, DC: World Bank.

PART IV

Extraneous Influences and Ethics in AI Applications

CHAPTER 17

Prevent AI from Influences: A Challenge for Lazy, Profligate Governments

Amita Singh

Disaster Governance: Managing Mechanized Humans or Humanized Robots?

The world is entering a dark tunnel of new technology missions in disaster risk reduction (DRR). The strongest challenge to be confronted by disaster governance in times to come is from AI and yet the alert is so feeble. The world of decision making in organizations both public and private would now onwards be constituted not only of human beings but of artificial human beings or robots. We are already quite dependent upon AI which controls our lives as smart phones as well as our behavior. AI works

Some sections of this paper were originally published in 2018 under the title 'AI: *In search of Regulations & legal Personhood'* by UNESCO in the volume 'Human Decision Thoughts on AI' brought out by the UNESCO-Netexplo International University Network, Paris, editors; Brigitte Lasry and Hael Kobayashi. The author also has the copyright to re-publish.

A. Singh (✉)
Centre for the Study of Law, Governance and Special Centre for Disaster Research, Jawaharlal Nehru University, New Delhi, India

T. V. V. Kumar, K. Sud (eds.), *AI and Robotics in Disaster Studies*, Disaster Research and Management Series on the Global South,
https://doi.org/10.1007/978-981-15-4291-6_17

through an algorithmic regulation and therefore acquiring human imageries does not entitle them Legal Personhood even though these machines may be much more intelligent and rational than most human beings. The world of work, leisure and war is already much governed by these intelligent machines to regulate human behavior, markets and government in addition to sometimes also seeking compliance, improvising market behavior and managing international borders. This opens up a whole gamut of citizen's relationship with the state in which the old rules of the game may prove incapable of handling these multifarious new challenges. Yet in disaster management it has become indispensable to count on this technology. The total change is so highly supportive to disaster management but if left unregulated this may wipe out those it goes to rescue. Given the large majority of apathetic and incompetent political and administrative leaders at the helm of state- and district-level disaster management departments, all new technology adoptions do need a smart, humane and knowledgeable disaster leader to press on its button.

The fear is embedded in the nature of this technology. In 2015, more than a thousand experts including key robotics leaders in the industry like Elon Musk the CEO of the AI Company like Tesla and Space X, Steve Wozniak the Apple co-founder, Demis Hassabis CEO of Google DeepMind and Stephen Hawking the legendary author in science signed an open letter demanding a proactive regulation of AI as it posed an 'existentialist threat to humanity'.[1] Following this alert by leading leaders of the robotics industry, the European Parliament passed a resolution to regulate AI which drew a much desired political attention in the development of European Robotic Industry.[2]

What Is Likely to Go Wrong?

AI has been brought into the administration of tasks which make human life better in many ways, i.e., smart phones, medical diagnosis, big data, communicational learning, organizational management, productivity and profits. An intrusion into spheres of privacy is almost accepted for state management of larger public good as bank details, medical information, investments, income tax, credit and citizenship details are not in public sphere nonetheless not sacrosanct from extraction and scrutiny whenever needed. The buck does not stop here as AI is pushing military organizations such as the Pentagon into a robot race which can lead to a form of a holocaust like problems when the machine is given to decide what's best for a situation.

AI also has high potential support for those handicapped with physical, visual, gender or age related disability. The masking software technology which can mimic any human voice and can access an external memory to react and imitate in a given situation like a human brain have the potential of criminalizing the clueless public sphere of people challenged by age, education, literacy, disability and poverty. The Google acquired British AI Company DeepMind Technologies Ltd. designed a program that "mimics any human voice" and claims to be "reducing the gap with human performance by over 50 percent." Such machines interfere and influence[3] or even capture neurological activity of the human brain and intrude their privacy of data.

The perversion of the innocent and benign Microsoft Chatbot software called Tay is of special attention here. On 23rd March 2016 this software was invented with the purpose of improving Microsoft's understanding of conversational teen language. The software was supposed to learn and improvise as it talks to people. In less than a day the software shocked its creators with its learning skills and also the skill to improvise from those racist, misogynistic and anti-feminist during political mudslinging that it started tweeting in a deeply offensive and illegitimated expressions to attack and abuse similar to what it learnt and heard from people tweeting.[4] This indicated an urgent need for revisiting the harassment at workplace laws, liability laws and also punitive laws suggested in grievance redressal. Microsoft took it offline in a day.

There are some unseen influences of AI which are likely to obstruct resilience building of local communities and one such negative impact is upon the job market as the data says. The 2014 Deloitte and Oxford University study[5] suggests 35% of UK Jobs are at risk. The Indian job market may see artificial intelligence (AI) and automation as the new drivers of employment in the $150 billion information technology (IT) industry which now employs more than four million people. However, the threat of job losses due to AI has been put to as high as 69% in India and 77% in China. The Chief Technology Officer of IT Services of Wipro, a leading IT firm in India, K.R. Sanjiv, warns that if AI is not planned well and addressed holistically, it is a disaster in the making. He spoke during a discussion on the 'rise of machines and future of human labour' at the Confederation of Indian Industry (CII) in Bengaluru.[6] Altus Consulting[7] has further added to the concern by suggesting that as the Robo-Advice, which is an automated, process driven financial advice, develops avatars that will replace human assistance and this would affect the whole of the banking and retailing sectors.

AI can read facial expressions and body language. This data is stored and protected as personal data and the robot is programmed to compare and identify physical distinctions, impairments, gender and disability. As used in recruitment and selection processes this may bring to surface some of the conventional and legally prohibitive tendencies of administration, organizations and departments towards efficiency, ability and capacity in defiance of the laws of equal opportunity, non-discrimination and gender which form the basis of affirmative action in public opportunities and distributive justice.

In some situations AI can work in a manner which cannot be explained even by its creators. Two years ago Toyota had to call back[8] many of its cars in USA which did not follow the brake and speed instructions programmed in it, which led to fatal situations for some its buyers. Similarly Nvidia, a chip maker company's vehicle did not follow a single instruction provided by its engineer or programmer.[9] The Tesla autopilot crash in Montana, Pennsylvania is a similar accident due to the failure of AI.

The drone accidents are a reality in which the recognition and identification went wrong and bombing was done at wrong targets.[10] Pentagon is spending billions to repeat the former 1970s MAD (mutually assured destruction capability) race of the cold war era to a robot race by designing lethal robotic fighter jets and submarines that would go into combat and can decide what to attack, can improvise images and understanding of targets, recollect learning and memory through which they can hunt down enemy targets without any such support or command from humans. This starts a new arms race in the world in which China and Russia are not far behind to USA's defense budget of an astounding $18 billion for three years on autonomous weapons fitted with AI. The human soldier and in their presence, any possibility of compassion, dilution of violence and prospect for peace would not be part of the erstwhile human warfare which generated these outcomes in the midst of bloodshed. Yet the most demonic is the situation in which these autonomous weapons performing reconnaissance, identification and recognition can start targeting their own people and property or fail to stop killing as the war is called off. These accidents have happened and are no more figment of one's imagination. Termed as killer robots and terminators these machines are a major threat to mankind.

Nature of New Regulations Needed

Both during preparedness and when the disaster strikes one ought to apply AI fitted equipment and machines with great caution and after testing their repeated trials. AI applications should be a gap filling device for SDMAs and DDMAs with sharp focus upon accountability and transparency rather than just speed. A pilot study in a small disaster area should be undertaken and a wider interdisciplinary expert committee could make an assessment of its impact and capacity to make a difference. As this technology is already labelled as a no. 1 risk for this century it needs tough regulations. Companies like the Google have set up an internal ethics review board to oversee its work in artificial intelligence. The information is to ensure that projects are not abused. On a wider side robots could be considered for awarding some rights and also liabilities at some point of time. The purpose of such a regulation should be well understood as it would part ways from the regular regulatory rationale for technology. Robots are technology of our future henceforth demands a regulatory direction before the market becomes impossible to regulate (Fig. 17.1).

If a robot is cautioned or controlled by law, its programmer would be liable for culpability in the accidents or crimes which could follow. This can replicate the case for corporations or companies which are sued for wrongs committed by them. Some may disagree with the grant of

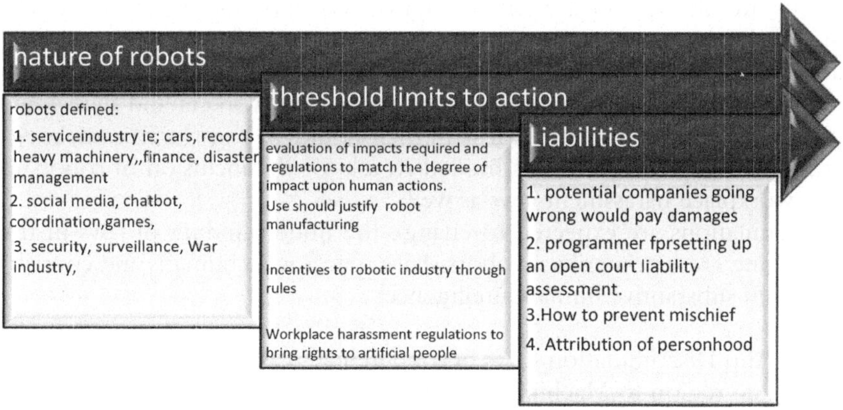

Fig. 17.1 Regulatory design

personhood but a learning robot may become compliant if the regulations are also programmed within the machine. These could best be ad hoc regulations only. The Resolution of European Union Resolution n. 59, let. E and F) is not opposed to personhood if that facilitates better monitoring and communication with stakeholders. This could be ensured by establishing an AI monitoring Expert Cell at NDMA in a bit faster pace than routine performance. Maybe outsourced to a third party with an established track record.

Robots may also be programmed to respect privacy and not to use a particular type of language, expression and behavior. This may equally apply to those who in mischief and fun corrupt a machine like what they did to Tay, the Microsoft Chatbot. Other service machines which go wrong could have a culpable human being as its programmer who failed the machine. Privacy regulation involves some of the biggest corporate crimes. Robots are used to steal company data or private data of users or clients to help other rival corporations.

Fixing liability is not that difficult if in robotics the human-machine interaction could be segregated to define clear cut roles. The victims of robotic failures are granted compensation and such funds be kept aside by each manufacturer at the inception of the robot. Compulsory insurance provisions for robotics should be enforced to address many risks likely to arise in practice.

The Constitution of any country and the UN Declaration of human rights Resolution on a right to dignity, equality and freedom of self-determination, as also suggested in the 68th General Assembly resolution of United Nations. Those working and communicating with robots at their workplaces generate respect for them as coworkers and detest any hateful or physical misbehavior with their bodies in angst or in casual negligence and realization. This has led to ask for the inclusion of robots in the workplace harassment laws as well.

Regulations are expected to change the understanding of laws in the following areas primarily as robots share spaces with humans and contribute with substantive human intelligence;

1. Anti Discrimination Laws beyond gender, caste, class, race, disability and region to Machines.
2. Labor laws for machines (national and International Labour Organization) which include workplace harassment laws, time and productivity management laws

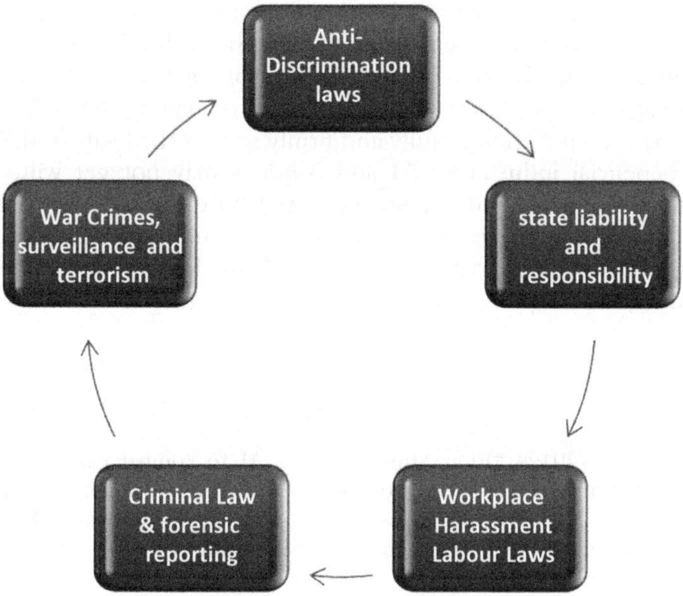

Fig. 17.2 Five areas of suggested legal changes

3. Criminal Liability Laws, forensic reporting, trafficking and cross-border crimes which may tag robots sometimes.
4. War crimes, surveillance and terrorism by humans and by machines.
5. State liability and responsibility in working with machines as humans (Fig. 17.2).

Conclusion

Regulating robots and AI applications in disaster management is a key concern for human safety, administrative accountability and future of mankind. Robots are becoming more intelligent and encompassing at workplaces. Some may be harmless companions to coworker human beings and may put in their greatest support to them but some may get corrupted and harm people or commit crimes such as frauds, theft and killing. Even though robots are a great panacea for all forms of disaster management, they need to be in the hands of their creator managed by an expert committee. Little work is being done in managing privacy and sustainable life and disaster management authorities should link the Disaster Management Act 2005 to the challenges of this new technology. This article suggests and draws attention towards regulating the robotic industry stringently but thoughtfully and firmly so that the fruits of this potentially beneficial industry of AI and robotics may not get wilted under slothful governance and corrupt outdated bureaucracy. The trial period should have already started by now but we are still debating and may hastily implant it without trials and then watch the dreadful impact becoming a shuttlecock of politics rather than a laboratory of transdisciplinary experts.

Notes

1. Gibbs, (2017) "Elon Musk: regulate AI to combat existentialist threat before its too late", Read more: http://www.dailymail.co.uk/sciencetech/article-2907069/Don-t-let-AI-jobs-kill-Stephen-Hawking-Elon-Musk-sign-open-letter-warning-robot-uprising.html#ixzz4swYzy6zb
2. Schafer, B (2016) 'Closing Pandora's Box?'
3. '*The Last AI Breakthrough Deep Mind Made Before Google Bought it for $ 400m, The end is nigh. Humans have lost another key battle in the war against computer domination*', 30 January 2014 blog, retrieved from https://medium.com/the-physics-arxiv-blog/the-last-ai-breakthrough-deepmind-made-before-google-bought-it-for-400m. on 15.09.2017.
4. Vincent, (2016) 'Twitter taught Microsoft's AI Chatbot to be racist asshole in less than a day' 7.
5. Deloitte Touche Tohmatsu Limited ("DTTL"), and Oxford University (2015) From Brawn to Brains.
6. Abrar (2017) 'Artificial Intelligence Imperils India Inc jobs'.
7. Altus Consulting White Paper (2014) 'Rise of the Machines: Where Next to Robo Advice' 7.
8. Boomey, (2016) Potentially deadly brake defect triggers Toyota Prius recall.
9. msmash from the problems-loom dept (2017) 'A Big Problem with AI: Even its Creators Can't Explain How it Works'. Tuesday April 11, 2017 @11:20AM, retrieved from https://apple.slashdot.org/story/17/04/11/1448202/ on 15.09.2017.
10. Rosenberg, (2016) 'The Pentagon's 'Terminator Conundrum': Robots That Could Kill on Their Own'.

References

Abrar, P. (2017, August 20). Artificial Intelligence Imperils India Inc jobs. *The Hindu*, Bengaluru, 20:55 IST. Retrieved from http://www.thehindu.com/business/Industry/artificial-intelligence-imperils-india-inc-jobs/article19529813.ece

Altus Consulting White Paper. (2014). *Rise of the Machines: Where Next to Robo Advice*. Retrieved from https://www.altus.co.uk/consulting/presentations-rise-of-the-machines-where-next-for-roboadvice/? On 15 Sept 2017.

Boomey, N. (2016). Potentially Deadly Brake Defect Triggers Toyota Prius Recall, *USA Today*, Published 4:23 am, ET Oct. 12, updated 3:05 pm ET Oct. 12.

Deloitte Touche Tohmatsu Limited ("DTTL"), and Oxford University. (2015). *From Brawn to Brains, The Impact of Technology on Jobs in the UK*. London: Creative Studios. Retrieved from https://www2.deloitte.com/content/dam/Deloitte/uk/Documents/Growth/deloitte-uk-insights-from-brawns-to-brain.pdf. On 15 Sept 2017.

Gibbs, S. (2017, July 17). *Elon Musk: Regulate AI to Combat Existentialist Threat Before Its Too Late*. 11:00 BST. Available at http://www.theguardian.com/technology/2017/1july/17/elon-musk-regulation-ai-combat-existential-threat-tesla-space-ceo

Rosenberg, M., & Markoff, J. (2016, October 25). *The Pentagon's 'Terminator Conundrum': Robots That Could Kill on Their Own*. Retrieved from https://mobile.nytimes.com/2016/10/26/us/pentagon-artificial-intelligence-terminator.html. On 15 Sept 2017.

Schafer, B. (2016). Closing Pandora's Box? The EU Proposal on the Regulation of Robots' Pandora's Box. *The Journal of the Justice and the Law Society of the University of Queensland, 19*, 55–68.

Vincent, J. (2016). *Twitter Taught Microsoft's AI Chatbot to Be Racist Asshole in Less Than a Day*. Retrieved from http://www.theverge.com/2016/3/24/11297050/tay-microsoft-chatbot-racist. On 15 Sept 2017.

CHAPTER 18

The Final Alert on Ethics in AI Based Technology

Vaishali Mamgain

Summing Up Ethical Questions Around AI

I structure questions along two strands which would be interwoven in this note. The first is an acknowledgment of the immense importance of artificial intelligence in disaster management and as a corollary I'd like to sound a cautionary note about the dizzying potential for poor governance, injustice and deep inequities.

An Example from the Chapters

As a way to highlight these concerns I refer to a scientist's use of GIS mapping to demonstrate that the flooding pattern of a river is predicted more accurately by models using machine learning (AI) rather than the forecasts made by conventional mathematical models. Being able to predict which bank will flood and at what rate is invaluable when deploying resources to

V. Mamgain (✉)
Bertha Crosley Ball Center for Compassion, University of Southern Maine, Portland, ME, USA
e-mail: mamgain@maine.edu

T. V. V. Kumar, K. Sud (eds.), *AI and Robotics in Disaster Studies*, Disaster Research and Management Series on the Global South,
https://doi.org/10.1007/978-981-15-4291-6_18

evacuate residents along a river. The improved precision that gives people or governments a few additional hours to react might save thousands of lives. This GIS based mapping of floods is cost-effective as well as strengthens human security. This is indeed wonderful! However, as an ethicist I want to question what happens *after* this forecast is made?

Information Is Power!

Ethical governance requires that we are scrupulous about how this data is disseminated.

As a society, we have to ensure that information about potential disasters is made available in a timely and transparent manner and not along established axes of power, privilege and wealth. Another confounding factor is the possibly fake information reported on social media platforms. This makes it even *more* critical that people are able to rely on their governments to give them timely and accurate information on a platform they trust. This is a core governance issue!

Unequal Access to Resources

If we can assume that information is made available in a timely manner, disaster response *has* to take into account peoples' unequal access to resources. For instance, when authorities predicted the ferocity of Hurricane Katrina in the United States, the people who evacuated were those who could afford to leave! These were people with some means- cars that worked, savings accounts and the ability to live in hotels or with relatives for an unspecified period.

In the 2019 fires that ravaged Los Angeles' most affluent neighborhoods the wealthy residents evacuated without letting their staff—their housekeepers, gardeners, chauffeurs—know their decisions. Entire service crews arrived *into* the most devastating fires Los Angeles has known in recent times. No matter the technology if our communities have failed to develop a culture of care for all beings we cannot flourish.

Inequality Between Countries

The previous examples referenced grave inequalities within a country but we have to acknowledge the inequality in and between countries also needs to be addressed. People in the global South will disproportionately bear the impact of climatic disasters created by fossil fuel consumption by the global North. Within the South the communities that will be most affected are those with the fewest economic resources. While we can celebrate early warning systems made possible by AI and funneling resources into such technology is important, if it is not accompanied by a commitment to helping displaced persons rebuild their lives and livelihood AI's marvels are a pyrrhic victory indeed.

The Benefits/Dangers of Cell Phone Data

The use of personal cell phone data might be a way to assist and rescue missing people during disasters; one researcher presented the overwhelming advantages of smart cities in the context of disaster management and an Israeli researcher made a powerful case for the use of facial recognition software to aid rescue personnel. On the face of it, it *is* tempting to imagine that this data will be used only in the event of a disaster but can we really believe that governments or anyone with access will be benevolent? Most countries have weak privacy laws and even those with stricter laws require an informed citizenry to assess how and in what circumstances their information is used. China's use of facial recognition software to surveil and target Uighur Muslims is not a farfetched sci-fi fantasy but a well-documented fact. San Francisco, aware of the potent power of this technology has banned the use of facial recognition software by local government including law enforcement agencies.

Consequences and Motivation

How then do we align our advances in artificial intelligence with good governance and ethical conduct? The answer is not easy in a dynamic world! A few years ago, on a flight from Boston a young post-doctoral student from the Massachusetts Institute of Technology was seated next to me. He told me he had been working on drones and confessed being deeply concerned about the uses of the technologies he had/was working on. "I am responsible," he said, "for any effects my inventions have." Karma!

His Holiness the 14th Dalai Lama was once asked by scientists whether it was okay to experiment on animals. Responding after much thought he said: although there were negative effects from killing an animal one might also consider whether that action might have a positive impact in the world. Then, His Holiness said, "The other consideration is what is in your heart as you do the experiment? What is your deep motivation?"

Guideposts as We Move Ahead

I invite us to consider this query a touchstone: what is my motivation? Whether as a scientist, a legal scholar, an emergency medical professional or an army officer trained in disaster response, do we know what our motivations are? If we have a clear intention to help sentient beings and we understand the *complete and sometimes complex* consequences of our actions, it can help us be humble and honest as we navigate this complexity.

This work on an individual level has to be buttressed by our commitment to systemic change to bring forth a society with a more just distribution of resources. Then and only then, can our communities share the benefits of better technologies in an equitable way. Dr. Cornel West famously said, "Never forget justice is what love looks like in public." When our institutional responses reflect this sense of justice we will build a more loving and kind world. Then, we hope, the conversation can shift from managing disasters to recovery, restoration and flourishing!

Conclusion

AI has no boundaries and accountability may run into several layers of technology manufacture. AI multinational companies such as Alphabet, Amazon, WhatsApp, Facebook, IBM and Microsoft are pervasive. They have users across the economic and social divides. Its ubiquitous presence has made it indispensable to debate on sound and focused usage of this new technology in disaster management where one comes to test its potential only when disaster strikes even though it could have done much at an earlier stages of preparedness. Ethics has worried thinkers like Stephen Hawking and Elon Musk who signed an open letter cautioning against pitfalls which are likely to be encountered on the usage of AI. While AI can do wonders on identifying risks and planning evacuation in a cost effective manner it can create a major disaster as well if left uncontrolled. The letter, titled **"Research Priorities for Robust and Beneficial Artificial Intelligence: An Open Letter"** which mentions, "The development of systems that embody significant amounts of intelligence and autonomy leads to important legal and ethical questions whose answers affect both producers and consumers of AI technology. These questions span law, public policy, professional ethics, and philosophical ethics, and will require expertise from computer scientists, legal experts, political scientists, and ethicists" (2015, p. 107), is an indication that AI should not be left in the hands of those most likely to misuse it under business or political compulsions.

Reference

Russell, S., Dewey, D., & Tegmark, M. (2015). Research Priorities for Robust and Beneficial Artificial Intelligence. *AI Magazine, 36*, 105–114. Association for the Advancement of Artificial Intelligence.

The manufacturer's authorised representative in the EU is Springer Nature Customer Service Centre GmbH, Europaplatz 3, 69115 Heidelberg, Germany. If you have any concerns regarding our products, please contact ProductSafety@springernature.com

Printed and bound by CPI Group (UK) Ltd, Croydon, CR0 4YY
30/09/2026
02225461-0009